WRITING
FOR THE
COMPUTER INDUSTRY

WRITING
FOR THE
COMPUTER INDUSTRY

KRISTIN R. WOOLEVER
Northeastern University

HELEN M. LOEB

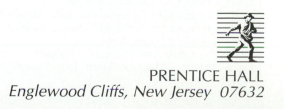

PRENTICE HALL
Englewood Cliffs, New Jersey 07632

Library of Congress Cataloging-in-Publication Data
Woolever, Kristin R.
  Writing for the computer industry / Kristin R. Woolever, Helen M.
Loeb.
    p.   cm.
  Indludes index.
  ISBN 0-13-971227-5
  1. Electronic data processing documentation. 2. Technical
writing. I. Loeb, Helen M. II. Title.
QA76.9.D6W66 1994
808'.066004--dc20   93-20056
                                                     CIP

PRODUCTION EDITOR: *Kerry Reardon*
ACQUISITIONS EDITOR: *Alison Reeves*
COVER DESIGNER: *Terrapin Graphics*
MANUFACTURING BUYER: *Herb Klein*
EDITORIAL ASSISTANT: *Kara Hado*

 © 1994 Prentice-Hall, Inc.
A Paramount Communications Company
Englewood Cliffs, New Jersey 07632

Printed in the United States of America

10 9 8 7 6 5 4 3 2 1

ISBN 0-13-971227-5

PRENTICE-HALL INTERNATIONAL (UK) LIMITED, *London*
PRENTICE-HALL OF AUSTRALIA PTY. LIMITED, *Sydney*
PRENTICE-HALL CANADA INC., *Toronto*
PRENTICE-HALL HISPANOAMERICANA, S.A., *Mexico*
PRENTICE-HALL OF INDIA PRIVATE LIMITED, *New Delhi*
PRENTICE-HALL OF JAPAN, *Tokyo*
SIMON & SCHUSTER ASIA PTE. LTD., *Singapore*
EDITORA PRENTICE-HALL DO BRASIL, LTDA., *Rio de Janeiro*

CONTENTS

PREFACE

The computer industry's need for technical writers and editors has developed and expanded almost as rapidly as its technology. Concurrent with the increase in computer sales to home and small business users and the growth of computer applications is the growing need for documentation and for people trained to write it. Companies have great expectations for their documentation professionals. More often than not, technical writers in the computer industry carry the burden of explaining the hardware or software, instructing the user, and contributing to the company's marketing efforts.

In the recent past, technical writers who had any formal technical writing training used textbooks written to teach engineers to write reports. The technical writer learned about writing for the computer industry in several ways: in-house training programs, company folklore, or seat-of-the-pants experience. Now that the industry is maturing, companies want to hire people they do not have to spend a lot of time training—they expect writers to come to work already knowing what to do.

Writing For The Computer Industry is designed to help new technical writers and technical writing students understand the basics of writing computer documentation. Books about writing have been around for hundreds of years—rhetorics, readers, and grammar texts have guided writers since the 16th century at least. Today, books about computers are spilling off the shelves in bookstores and libraries—applications books, programming books, and buyers' guides. Yet even with all the attention to both writing and computers, the number of books that combine the two can be counted on one hand.

An understanding of advanced computer technology does not necessarily translate into effective documentation skills. And although a firm grasp of traditional writing techniques is essential for technical writers, effective documentation goes beyond basic writing skills and requires writers to be graphic designers, marketing specialists, researchers, and psychologists—as well as good writers. By focusing on how these varied skills work together, *Writing For The Computer Industry* offers a comprehensive view of writing computer documentation and gives practical advice about how to get the job done.

The book begins where communications specialists begin: analyzing audience and purpose. Then, in clearly defined steps, the book takes writers through the documentation process, from the preliminary planning to the final edits. Discussed in these pages are all aspects of the documentation cycle: getting information, working with a development team, writing hardcopy, developing online instructions, editing, testing, and establishing professional standards. Writers can use this book in different ways:

- as a classroom text
- as a reference book
- as a company training guide
- as a self-paced tutorial

Because the book is designed for both students and practitioners, the exercises within each chapter work in either university or industry settings. They serve as practical checkpoints along the way for writers to test their skills. Some people may apply the exercises to one specific piece of documentation, taking that one document through the various development stages. Others may decide to work on several smaller projects and apply the exercises to different documents. Either choice works well. Likewise, the text can serve as a resource for classroom assignments as well as for actual on-the-job documentation projects.

No matter how people use this text, *Writing For The Computer Industry* serves as a practical guide to the documentation process. In compact form, the book is a valuable toolkit for both students and practicing professionals in today's high tech environment.

Because writing any sort of a practical book is not a solitary process, many people have contributed to creating this text. Marguerite Krupp has read the manuscript with a keen editorial eye and has offered invaluable advice. Members of the Society for Technical Communication, Boston Chapter, have provided ideas, examples, and continued support. The following professionals have reviewed drafts and have made much-needed editorial comments: Thomas Warren, Oklahoma State University and Stuart Selber, Michigan Technological University.

I would like to dedicate this book to my friend, colleague, and co-author, Helen M. Loeb, who died before seeing this work in print. Her memory continues to inspire new generations of documentation specialists.

Most of all, this text is the result of continuing conversations with students past and present in Northeastern University's graduate technical writing programs. They are the real authors of this book. It belongs to them.

Kristin Woolever
Boston, Massachusetts

CHAPTER 1

TYPES OF COMPUTER DOCUMENTATION

When the computer industry was in its infancy, no one thought much about writing manuals to accompany the miraculous new machines. People who needed to operate these computers took special training sessions taught by experts and then passed along the knowledge to those who came after them. Later, as computers became more integrated into our daily lives and more available to the general public, companies began mailing instruction books with the equipment. These early computer manuals were usually written by engineers who were experts in computer technology but not so well-versed in methods of explanation. The resulting manuals were often bulky three-ring binders including everything from installation and maintenance procedures to operating instructions and sometimes even the mathematical theory behind the technology.

Times have changed. Today, computer documentation comes in all kinds, shapes, and sizes, depending on the purpose of the document. No longer is the writing task assigned solely to engineers and completed almost as an afterthought when the product is ready to ship. Instead, many companies have entire departments devoted to technical writing, and these writers are often involved in the development of the hardware or software from the very first planning stages. And, rather than shipping one bulky manual with the product, most companies ask their writers to design several smaller books, each serving a different purpose. These various books and other supportive materials make up the "documentation set" that accompanies the product, allowing the user to choose exactly the right manual for the task at hand.

What kind of material makes up these sets? That question opens a Pandora's box of computer documentation styles, types, and sizes. Beyond the basic classification of computer documentation into hardware and software categories, there are many subcategories that can seem confusing at first. Hardware documentation includes reference manuals, user manuals, installation and maintenance manuals, owners' manuals, and quick reference materials. Software documentation includes all of the above manuals, plus tutorials—and is further divided into the sub-subcategories of systems software documentation and applications software documentation. (Systems software enables parts of computer systems to work together. Applications software enables end users to perform specific tasks.) To add to the confusion, all of these categories (both hardware and software) can include online documentation as well as hardcopy manuals.

In spite of this maze of document classifications, technical writers can feel well prepared to construct all kinds of computer documentation if they understand the essential characteristics of five basic types: user guides, reference manuals, quick reference guides, tutorials, and online documentation. These five are the basis for all the others.

═══USER GUIDES

A user guide is a manual that gives step-by-step instructions for first-time users of a particular product. Because the writers of these guides assume their audience knows little about the product, the manuals begin with the most basic steps and move gradually toward the more advanced tasks. This "task-oriented" organization is usually chronological, arranged as closely as possible to the order in which the user will perform the procedures. For example, a typical user guide for a word-processing program might begin with a chapter called "Getting Started" which lists the features of the program and helps users understand how to load and run the software. The next chapter might be a section on creating documents, the third might introduce editing techniques, and so on. New users should be able to follow these steps in basically the same order they are presented. By the time users have finished the entire manual, they should have moved logically through all the functions of the software and understand how to use the product.

To give you a better idea of how this works in practice, Figure 1-1 shows a sample table of contents from a typical user guide. Note how each

Contents

ix Introducing Lotus Manuscript

xi Terms
xii About This Book
xii Where to Begin
xiii Keystroking Conventions

1-1 Session 1: Starting with Text

1-1 Starting Manuscript
1-2 Creating a New Document
1-4 The Manuscript Screen
1-5 The Keyboard
1-7 Entering Text
1-8 Correcting Mistakes
1-9 Using the Function Keys, Panels, Checklists, and Menus
1-12 Changing Text Attributes
1-16 Warning Messages and Help
1-18 Saving a Document
1-19 Draft Printing
1-19 Ending a Manuscript Session
1-20 Summary

2-1 Session 2: Working with Blocks

2-2 Retrieving an Existing Document
2-3 Examining the Blocks of Text
2-4 Appending a Block
2-5 Moving and Copying Blocks
2-5 Cutting a Block
2-6 Pasting a Block
2-8 Copying a Block
2-9 Using Move

iii

FIGURE 1-1 User Guide Table of Contents (*Source*: Lotus Manuscript is a registered trademark of Lotus Development Corporation. © 1990 Lotus Development Corporation. Used with Permission.)

section builds on the information given in the previous sections, organizing the procedures by chronological tasks. The users can progress gradually into more advanced functions.

▬▬REFERENCE MANUALS

Unlike user guides, reference manuals do not assume a novice user. Instead, the audience for this type of manual comprises people who already know enough about the product (or about computers in general) to operate the software or hardware but who need help with specific tasks. For example, assume you have bought a new word-processing software package and have spent a few days learning to use it by reading the user guide. The following week, you find that you need to transfer information from one file to another, but you can't remember exactly how to do that. Rather than wade through the step-by-step introduction to this process in the user guide, you want to consult a manual that gives refresher information about that one task.

Reference manuals allow users to find specific information quickly and are designed to present material in self-contained units so that users do not have to go elsewhere in the text to complete the task. Often these books are organized by commands, such as "copy" or "delete," or by other highly specific tasks, such as "splitting screens" or "modifying the default printer setup." In any case, for this organization to work, users must know what they want to do and understand the terminology that will allow them to do it. The manual spends little time with explanatory material.

A common practice is to combine user guides and reference manuals into one book. Writers can do this by dividing the book into two parts—the first containing introductory user material, and the second containing commands. Another way to combine the two is to make the user guide more "reference-able" in its design. Chapters 6 and 11 give tips on how to design manuals that serve as easy reference.

Figure 1-2 shows a sample table of contents from a reference manual for the same word-processing software used in the earlier example. Note the difference in organization and in orientation.

▬▬QUICK REFERENCE GUIDES

Quick references assume an even more knowledgeable user. In fact, the audience for quick reference guides are people who know the product well and need only brief reminders of how it operates.

These references are often so accessible that users don't even have to turn pages—the information may be printed on a card that fits into a shirt pocket or on templates that fit over the computer keyboards. Even when the material requires more than one page, the total number of pages is minimal and the physical document is usually designed so that it can be easily accessible on the job. It is rarely attached to a larger manual.

Contents

v

FIGURE 1-2   Reference Manual Table of Contents (*Source*: Lotus Manuscript is a registered trademark of Lotus Development Corporation. © 1990 Lotus Development Corporation. Used with permission.)

As you might expect, quick references are nothing more than thumbnail sketches of how to use the product. There is no explanatory information at all, nor are there features such as a table of contents or an index. These guides really are nothing more than memory joggers designed to be used at a glance, as the quick reference card shown in Figure 1-3.

TUTORIALS

Tutorials, on the other hand, contain a great deal of explanatory material and require the users' full attention. As the name suggests, these documents are teaching tools that provide users with mini-lessons on how to use the product. In some cases, the tutorials are presented on screen, allowing the user to interact with the computer, much as students do with their classroom teachers. The tutorial program may ask you to type something or use the mouse to perform a task, and then the program gives you immediate feedback on your performance. It may beep at you, flash messages on the screen, or even present video sequences indicating praise or criticism of your work. Whatever interface the tutorial uses, it always requires you to participate in the learning process.

Hardcopy tutorials (tutorials presented on paper rather than on screen) are also interactive to a certain degree. Organized in a series of "lessons," the tutorial asks you to perform various sample tasks that give you a sense of how the product works. Each lesson builds on the previous one and often asks you to add or delete something from the work you have completed in the previous lesson. As in the classroom, you should be able to finish the work in one sitting and have some definite feedback on how well you've learned the material. In hardcopy tutorials, that feedback may come in the form of a printout of the task you've completed, or it may come just in the satisfaction of getting the product to do what it was supposed to do. Either way, the instructions should indicate clearly how to measure your success.

Because they are designed to teach users specific tasks, tutorials are organized around actions, not system commands. This differentiates them from reference guides. For example, a tutorial chapter entitled "Lesson One: Create a Letter" tells you exactly what task you will learn and what tangible item you will have as a result of this lesson.

Tutorials are different from user guides because they present material in small steps, provide exercises, and reinforce through repetition. They are also more interactive than user or reference guides and often contain more "user-friendly" techniques such as an informal tone, extended metaphors, pictures, reassurances, and other similar teaching tools. Throughout a tutorial, you always have the sense that you are being led gently through the process by a friendly teacher who is infinitely patient. This learning environment works well for people who are unsure of their computer knowl-

FIGURE 1-3 Quick Reference Card (Used by permission of XyQuest.)

REFERENCE CARD
XYWRITE WORD PROCESSOR

Chapter number references on this card refer to the *Reference Guide*.

KEYS

Some of the most commonly used keys are listed here.

HELP *(Chapter 6)*

| Alt | F9 | | Help |

CURSOR KEYS – UP AND DOWN *(Chapter 3)*

| ↑ | ↓ | | Move One Line Up, Down |
| Ctrl ↑, Ctrl ↓ | | Scroll Text One Line Up, Down |

CURSOR KEYS – LEFT AND RIGHT *(Chapter 3)*

←	→		Move One Character Left, Right
Alt ←, Alt →		Move One Word Left, Right
Ctrl ←, Ctrl →		Move to Left End of Line, Right

TAB KEYS *(Chapter 3)*

Alt, Tab		Display Tab Menu
Ctrl Tab		Move Cursor to the Next Tab
Tab		Move Cursor and Text to Next Tab

COMMAND LINE *(Chapter 3)*

F5		Clear the Command Line
F9		Execute Command on Command Line
F10		Switch Between Command Line/Text

DEFINE KEYS *(Chapter 3)*

F1		Begin/End Block Define
Alt F1		Begin Column Define
F2 #		Save Defined Block
F3		Release Defined Block
F4		Define Line
Ctrl F4		Define Sentence
Shift F4		Define Paragraph
Alt F4		Define Word
Alt F5		Delete Defined Block
F7		Copy Defined Block
F8		Move Defined Block

EXPANDED DISPLAY *(Chapter 3)*

| Ctrl F9 | | Switch Between Normal/Expanded |

PAGE NUMBER DISPLAY *(Chapter 3)*

| Shift F9 | | Turn on Page-Line (P-L) Indicator |

WINDOWS *(Chapter 3)*

Ctrl F10		Display Window Menu
Alt F10		Switch Between Last Two Windows
Shift F10		Switch Through All Windows
Shift Ctrl n		Switch to Window n

Ruler markings: 0, 1, 2, 3, 4, 5, 6, 7, 8, 9, 1.0 inch, 1.5, 2.0, 2.5, 3.0, 3.5, 4.0, 4.5, 5.0, 5.5, 6.0, 6.5

Creating a 3-D Stacked Bar Chart

In a stacked bar chart, the bars are composed of segments that represent the variables. In this sample chart, you'll see total revenue for each
of the six months, as well as how each group performed in that month.

1. Click the **Chart** button.

2. Click the bar chart icon with the right mouse button.

   The Bar Chart Options dialog box appears. The options on the
   left side of the dialog box are covered in Chapter 4, "Numeric
   Charts." The chart styles are shown on the right side. You can
   select from (in order) overlapping bars, clustered bars (the
   default), or stacked bars.

3. Click the stacked bar icon.

4. Click **OK** and then **Redraw**.

   Notice that APPLAUSE adjusted the y-axis scale maximum to
   accommodate the largest cumulative value. To reduce the number of labels on the y-axis and the number of grid lines, you can
   increase the increment again. Just repeat the steps you followed earlier in this lesson to adjust the y-axis.

   If you change the increment to 2500, the chart looks like the
   one in Figure 2–7.

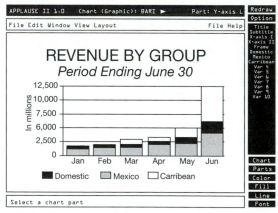

Figure 2–7 Stacked bar with y-axis increment of 2500

2–10   U

FIGURE 1–4 Tutorial Sample

Next add a 3–D effect to add an extra dimension to the chart.

5. Select the **Layout** menu and choose **3–D effect**.

   The 3–D Options dialog box appears. For this exercise, just select a 3–D view icon.

6. Click the upper left icon and click **OK**.

   The 3–D stacked bar chart appears as shown in Figure 2–8.

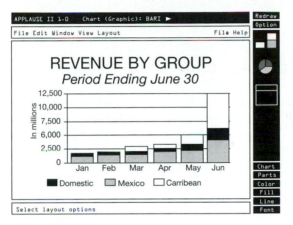

Figure 2–8 3–D stacked bar chart

Saving the 3–D Chart

To save the 3–D variation of BAR1, use the Save as command on the File menu. Save as is useful when you want to save different versions of the same chart.

1. Select the **File** menu and choose **Save as**.

2. Click the **Filename** text entry box and type **bar2**.

3. Click **OK**.

Now you can continue to the next lesson, and build a pie chart.

edge and for people who learn by doing rather than by reading. It's easy to plunge right in to a tutorial without having to read a lot of preliminary material or be distracted by explanations about things other than the discrete lesson at hand.

But tutorials do not serve well once you have mastered the product. If you attempt to return to the tutorial for help after you have used the product successfully a few times, you will likely find the pace too slow. Tutorials are effective only for the novice user; they are inappropriate for anyone with more experience with the product.

Figure 1-4 shows an excerpt from a tutorial. Note the informal tone and the slow, interactive nature of the material.

═══ONLINE DOCUMENTATION

Some people believe that hardcopy documentation will soon be a thing of the past and all instructions will be electronic or "online" (on the computer screen). Others believe that there will always be a mix of paper and electronic documentation, but everyone agrees that users will come to expect increasingly sophisticated online instructions. The advantages are obvious. As you are working, you don't have to stop and leaf through a manual to get help. Instead, you simply press a few keys, point to or touch an on-screen icon, and all the information you need appears on the screen. Although the information is brief and may not provide you with overviews or allow you to understand the workings of the whole product, it does allow you to get the help you need quickly.

There are many kinds of online documentation: error messages, help screens, online tutorials, dialogue boxes, and even the large category of hypertext-based electronic documents, which includes many on-screen links and user cues. Chapter 14 discusses online documentation in detail, but it's important to know at this point that the online medium affects the way the writer presents the information and the way the reader perceives and uses it. And, the medium offers the added potential for sound and animation. Any type of information presented on the screen follows unique rules that govern the way users process visual material. These processes differ extensively from the way users view the printed page, and designers of effective online documentation must allow for these differences. Figure 1-5 shows a simple example of an online screen.

═══FINAL THOUGHTS

Nearly all computer documentation falls into one of these general categories. Before you begin to write, you must take into account the user's

CM
PRMPT
INDEX WORD KEY ASCII ERROR HELP COMMANDS FUNCTIONS TEMPLATE DOS

About Save/Gets

You can have text blocks available for immediate recall with a single keystroke.

To do this:
1. Define the block of text you want to "save."
2. With the block highlighted, press F2.
3. Next press any letter or number (a–z, 0–9).
4. Press F3 to release the defined block.

Insert these blocks of text "Saved" using ALT and the letter or number used to save it. This is the "Get" part.

You can have up to 36 Save/Gets.

STSGT—Saves all of Save/Gets as a file.
LDSGT—Loads Save/Gets from a previously saved file.
CLRSGT—Clears all Save/Gets in memory.

FIGURE 1-5   Sample Online Documentation Screen

needs and the nature of the product in determining the type of manual you're composing. Then you won't be tempted to mix techniques and produce a patchwork document that includes a little bit of every variety but fails to serve the user well.

EXERCISES

1. Find three different computer manuals and analyze their basic characteristics. Then, categorize them according to the five types explained in this chapter. Do some of them fit in more than one place? Do some of them claim to be something other than what you think they are? Are they successful in their appropriate categories? Why or why not?
2. Find a relatively short user guide that you think is appropriately categorized. Look at the table of contents. How would you change that table of contents if the book were a reference manual instead of a user guide?

CHAPTER 2

DEFINING AUDIENCE AND OBJECTIVES

Understanding the specific audience and purpose for each book is perhaps the most important element in writing computer documentation. If writers don't do a careful analysis of these things *before* they begin writing document plans, schedules, outlines, and drafts, they may take an incorrect approach and consequently waste a lot of time. No matter how rushed you may be to meet a deadline, the time spent defining your audience and objectives for writing is essential to streamlining the whole writing process and to producing an effective document.

DEFINING AUDIENCE

Writers need a logical approach to determining exactly who their audience is. Until you have clearly defined your audience, you cannot begin the other preliminary research necessary to understand the product you're writing about, including interviews with the developers, studies of competing manuals, and discussions with technical support personnel. Why? The questions you may ask one audience are not the same questions you would ask a different audience, nor are the answers you receive the same. All the variables in your writing project depend to one degree or another on the audience. Consequently, you need to begin with as detailed a description of your audience as is possible.

How are you going to get this description? The first step is to understand the common types of audiences manuals are written for.

═══COMMON AUDIENCES FOR COMPUTER DOCUMENTATION

The people who read computer documentation are as divergent as the spectrum of people who work for a sprawling, multinational corporation. They range from company presidents to maintenance personnel, from research engineers to data-entry clerks. In addition, the booming home-computer market ranges from highly sophisticated adults who use computers to run home businesses or to figure their finances to small children who use computers for school projects or for the many software games available. These people's needs, interests, and goals are widely divergent in the way they approach computer documentation. As a result, computer documentation is written, organized, illustrated, and printed to meet various audience's needs.

In general, computer companies produce manuals for two broad groups: customers and employees.

CUSTOMERS

Customers may be divided into three subgroups:

- Those who can make or recommend a purchase agreement—executives, consultants, data-processing managers, and computer-savvy business people.
- Those who use the equipment—systems operators, workers who use computers regularly to perform their jobs, computer system maintenance personnel, executives and managers and others who may use a terminal or a PC only occasionally, and home-computer owners.
- Those who program the computers—programmers, systems analysts, and data-processing managers.

Each of these groups has individual needs and will use the documentation differently from the way other groups may use it. Therefore, technical writers assigned to write manuals for one of these categories should take into account these differences and design the manuals to fit the users' needs. The following descriptions of each of these groups may help writers to better understand what they need.

Those Who Can Make or Recommend a Purchase Agreement

In a corporation, final purchase or lease agreements are generally made by *executives or department managers* rather than the people who will actually use the piece of equipment. These executives may have the power to approve a purchase on their own, or they may meet with company comptrollers or similar personnel who are responsible for budget decisions. Their primary concerns are efficiency, cost, and company profits.

An outside *consultant* may also make a purchase recommendation after meeting with employees of a company, assessing processes that could be computerized (such as payroll, inventory, accounts payable, and so on), and recommending a computer system.

A *data-processing manager* who oversees all processes that will be computerized or will be transferred from one computer system to another will make a purchase recommendation based on his or her understanding of how closely, accurately, and efficiently the proposed system can process the company's information.

Finally, *computer-savvy business people* who have authority for their department's budget often recommend product purchases, if in fact they don't make the purchase themselves. These people combine their knowledge of business with knowledge of computer technology and look for products that will provide clear and significant benefits to their organizations.

Those Who Use the Equipment

This group is extremely varied and ranges from personnel whose primary responsibility is computer operations, to people who rarely use computers, to those who have purchased personal computers for home use.

Systems operators and systems managers are people who work at a "system console"—that is, a terminal that can control the flow of work through the computer. These operators are also responsible for troubleshooting, assuring security, and verifying that all systems and personnel are working together properly. They are the people that others in the organization turn to when things go wrong. They need detailed, specific information—usually in a hurry. They may also program computers and may perform minor hardware repairs.

Workers who use computers regularly to perform their jobs may be responsible for entering information into the system (the personal computers are usually linked in a "network" that enables these machines to share information), running procedures or performing specified jobs, and passing the computer's "output" along to a third party. Others may have jobs in which they use the computer as a tool, not as an end in itself. If the documentation gets in the way, they probably will not use the product. These people include data-entry personnel, word processors, checkout clerks in retail stores, bank tellers, medical personnel, scientific researchers, financial analysts, airline ticket agents, and warehouse personnel. They usually have minimal exposure to programming or to any computerized operations other than the entry of figures, words, or codes into their terminals. Workers who use personal computers include people such as newspaper reporters who work privately on their own computers but send and receive information electronically. People in this group may use some programming skills to customize their software to complete more efficiently the tasks they need to do.

Computer systems maintenance personnel are responsible for uncrating and setting up the hardware. They, too, have manuals written for them.

Executives and managers and others who may use a terminal or a PC only occasionally usually have computers in their offices to send messages, schedule meetings, and manage financial data, or to prepare drafts of memoranda or other reports. They may go days without using their computers and so may prefer programs and documentation that make it easy to pick up where they left off.

Home-computer owners are a fast-growing group of users. They may simply use "canned" applications, such as packages for balancing a checkbook, writing letters, or playing games, or they may become sophisticated programmers who tinker with existing software to make it more efficient.

Those Who Program the Computers

Programmers write software for computer systems and personal computers. They may be employed by the customer or may be "third-party" programmers who work on a consulting basis to write programs tailored to a customer's needs.

Systems analysts are people with extensive programming knowledge who evaluate a user's computer needs and the computer's capabilities, and plan operating methods and procedures so that the system can achieve the user's goals. Often they are also the ones who design the programs that the programmers construct.

Programmers and systems analysts are the audience for the bulk of computer reference manuals. Typically, they use documentation for systems software; programming languages (C, COBOL, FORTRAN, Pascal, and so on); utilities (programs that perform highly specific functions, such as sorting data, retrieving information inadvertently lost, and so on); computer applications manuals; and special programmer or systems analyst guides. They are also the audiences of special books, such as conversion manuals, or books that allow an analyst to decide how much storage space or main memory to dedicate to a particular set of files or set of instructions.

The goal of these manuals is to be as precise and complete as possible so that programmers will know virtually everything possible about the language, software, or utility in question, or about the capabilities of a computer system.

Finally, *data-processing managers* may do some programming as part of their jobs to keep the work flowing smoothly through the system. In small operations, they may in fact be systems operators, systems analysts, and/or programmers.

EMPLOYEES

The second of the two broad groups of computer documentation audiences is employees. This group, too, can be divided into subcategories:

- Sales and marketing representatives
- Systems engineers
- Field engineers
- Trainers
- Development engineers and programmers
- Technical writers

Sales and Marketing Representatives

Salespeople sell computer systems. They must not only know the capabilities of the hardware and/or software and how to make them work, but also how they can fit the users' needs. They must be as familiar as possible with the equipment in order to answer potential customers' questions and to recommend the size and complexity of computer appropriate for the buyers. Marketing representatives, on the other hand, are rarely directly responsible for sales. Instead, they identify target markets, project sales revenues, and may help design the product. In many companies, they may assist in writing sales brochures, along with the cost sheets and sales specifications. Their key need is documentation that identifies the features, benefits, and specifications of the product.

Systems Engineers

Systems engineers (sometimes called "sales" or "customer support" representatives) assist customers with programming and operating problems. They assist customers in converting manual operations to a computerized system or in converting from one system to another, and they are often present when the new system is installed. They use customer operating and programming manuals for reference, and may also write customized applications.

Field Engineers

These engineers install, maintain, and repair computer hardware. Although they work with customers, field engineers are employed by the computer company, and they have manuals written exclusively for them. These manuals explain circuitry, program logic, troubleshooting procedures, and other information that is proprietary to the company. Typically, these manuals are restricted to employees of the company so that customers cannot release the design information to a competitor, damage the hardware and thus void warranties or guarantees, or injure themselves.

Trainers

Trainers conduct educational sessions for new customers or customers with new applications. Trainers use customer documents as well as special training manuals designed specifically for their classes.

Development Engineers, Development Programmers, and Technical Writers

Within the computer company, these three groups of employees use computer documentation but are rarely listed as primary audiences. All of them use manuals for reference, both to verify information about the system and to provide samples for specifications or other new documentation they may be working on. They also use tutorials and user guides to become proficient users of new products.

ANALYZING THESE AUDIENCES

Now that you understand the general audiences common to computer documentation, you need to go beyond these general categories and determine even more specific characteristics of the people who will use the manual you plan to write. Within each of the preceding groups, there can still be a wide variety of users, and the nature of these groups varies also with the culture of individual companies. For example, at IBM, the work environment may be more formal than at a small start-up company located in someone's home. Every company has its own personality which is reflected in the way the company's computer users read the documentation. By paying attention to the psychological factors that affect the users' comprehension capabilities, you can write manuals that appeal more directly to your audience. The more you know about the *specific* users, the more readable and useful your documentation will be.

To help get a clearer sense of audience, you may find it helpful to answer the series of questions listed in Figure 2-1, the Primary Audience Worksheet. (If you have the opportunity, it's a good idea to pose these

---

AUDIENCE WORKSHEET
Primary Audience

1. What is the job function of the primary audience?
2. How will this audience use the document?
3. What is the educational level of this audience?
4. How experienced are the members of this audience in their jobs?
5. How experienced are they with computers? With your product?
6. What is their work environment like?
7. What is their interest level?
8. What biases, preferences, or expectations might they have?
9. With what other computer documentation are they familiar?
10. How much theory or "nice-to-know" information do they want?

FIGURE 2-1 Primary Audience Worksheet

questions directly to the users.) You can copy the worksheet and fill in the answers, or you may want to use it as a guide for composing your own list of questions. Whichever option you choose, make sure to ask questions that are specific enough to get a detailed picture of your audience's professional and psychological requirements. The next section contains a complete discussion of the ten questions on the worksheet and the rationale behind them.

Discussion of Primary Audience Worksheet Questions

1. What is the job function of your audience? The person who will use your manual probably belongs to one of the common audience types described previously. Once you determine what the job function is, you can make some general assumptions about the user's needs, based on the typical characteristics of that audience type. For example, your audience may be executives at an architectural firm, but because the firm is small, they would use the equipment to do some programming and data entry in addition to general work. Your manual should provide information they can use to do all of these tasks.

2. How will the audience use the documentation? These answers will give you more specifics about the user than Question 1 can do.

3. What is the educational level of the audience? Is your audience college-educated, perhaps with advanced degrees? Or, do they have two-year-college degrees or vocational-school training? The answer to this question gives you insight into your audience's reading level. The language, sentence lengths, and style you choose will differ depending on this answer. Remember that highly sophisticated and educated

specialists in one field often lack the skills, vocabulary, and interest in another field (such as computer technology).

4. and 5. What is the audience's experience? You need separate answers to two questions: You need to know how experienced the users are in their jobs and how experienced they are with the computer product you are documenting. An architect with twenty years of experience may be an expert at structural design but be a novice computer user. To address such a user, your document must use a sophisticated language about the product's architecture applications, but it must also present the computer instructions at a fairly basic level.

6. What is the audience's work environment like? Will the users be working in a noisy environment with frequent interruptions? Or will they have enough peace and quiet to really study the manual and read it in depth? The manual's design and level of detail must match the audience's ability to concentrate on the information presented.

7. What is the audience's interest level? Some audiences may be interested in the product to a point, but they are busy people. They will be in no mood to read extensive prose material—they want to get straight to the operating instructions. Other audiences may prefer to read contextual explanations. Still other audiences may be computer-phobic or genuinely hostile to the product; these users don't even want to open the book. This question really addresses reader psychology. Your organization, tone, and format will change depending on the user's interest in what you write.

8. What are the audience's preferences, biases, and expectations? This question is often more difficult to answer than Question 7, but it, too, addresses reader psychology. You may discover that your audience is highly interested in programming languages, but has a bias against English-like languages and toward assembly language. It helps to know these preferences in advance, so you can tailor the documentation to better meet user expectations.

9. With what other computer documentation might the audience be familiar? People tend to approach tasks in ways that are already familiar to them. If the other documentation works well or is part of the same instruction set, the users will be grateful to recognize some consistency in design and approach. If the other documentation is poor, you can benefit from knowing what doesn't work well and can design your documentation to correct these faults. The bottom line is that you should be familiar with the same books or online instructions that the users are, so that you can better understand their point of view. (Another note—if you're working on a document set with other writers, be sure to coordinate and standardize your approach, terminology, design, and so forth.)

10. How much theory does your audience want to know? This question asks the difference between the "need-to-know" and the "nice-to-know" information. It relates to your audience's job function, experience, and interest level. An experienced developer may want detailed theoretical information, while a data-entry operator may want none.

A WORD ON MULTIPLE AUDIENCES

Although your documentation will be shaped by your primary audience, you also need to have some information on other users. It is rare that a computer manual has only one audience, and you need to determine the significance of your secondary set of users. Why will these people use the book? What information will they need? Will their needs be compatible with your primary audience, or in conflict? It may help to fill out a worksheet for the secondary users after you have completed analyzing your primary audience (See Figure 2-2).

No matter what information you discover about your secondary readers, do not let this material dominate the book, particularly if there is a conflict between the user groups. For example, do not rewrite the entire manual at an advanced reading level simply because one or two sophisticated users might pick it up. But you may want to design your manual to accommodate two or more audiences. If your primary audience needs a user guide, but the secondary readers prefer a reference manual, you can design the book to incorporate both. If you are writing a document that will be used initially for training and later for quick reference, design it so that the summaries can be located quickly.

As a last alternative, you may find that you have a conflicting audience that is split down the middle—half developers and half data-entry operators, for example. The answer to this dilemma is to write two different documents, not one that tries to be all things to all people.

FIGURE 2-2 Secondary Audience Worksheet

AUDIENCE WORKSHEET
Secondary Audience

1. What is the job function of the secondary audience?
2. Why will this audience use the document?
3. What is the educational level of this audience?
4. How experienced are the members of this audience in their jobs?
5. How experienced are they with computers? With your product?
6. What is their work environment like?
7. With what other computer documentation are they familiar?

RESOURCES FOR DEFINING AUDIENCES

Even when technical writers know all the audience analysis questions to ask, they are left with a problem if they don't know where to go for the answers. If you work in a corporate environment, you have several resources within the company.

INTERNAL RESOURCES

First, ask your supervisor or project leader. In small companies, the supervisors may be able to describe the users of the product in detail, whereas in larger companies—especially multinational ones—assessment of audiences may be more difficult. However, most project leaders will have access to a marketing document variously known as a "project objectives statement," an "overview functional specification," or some similar name that describes the audience, though not in detail.

Other people at work may be able to help you. Your company may have a human factors department whose business is to understand the needs of various audiences. Make an appointment to talk with them about the product and the intended audience for it. Another source is the customer training department. If anyone knows the users' needs, fears, background, and experience, it is the trainer who teaches them how to use the products. Chances are that the trainers have also used other documentation your company has produced, and they can give you both an audience definition and some commentary on the strengths and weakness of those other documents.

Systems engineers, whose job it is to solve software problems for customers, and field engineers, who take care of hardware problems, are another reliable source of audience information. They will be able to tell you a great deal about the customers' environment. The other people you work with may also know something about your audience. Try customer support for firsthand knowledge of the problems users have with current manuals and products. These people, who answer telephone inquiries all day, certainly can tell you in great detail about the users they talk to.

Finally, you can speak to a variety of marketing people. The marketing department is responsible for market analyses and surveys and is a storehouse of excellent information. In some cases, the marketing people go out into the field with the sales people to sell the company's product, and in other instances, they work primarily to develop marketing strategies. Your wisest course is to find people who have been in the field recently, rather than to talk to those who haven't seen a customer at work for several years.

You can also find out something about your audience from previous documents your company has produced. Look at these manuals to see what kind of audience decisions other writers have made; then, compare this information with what you have learned in your own research. You may find these documents extremely helpful or extremely dated in their approaches.

EXTERNAL RESOURCES

After you have gathered as much information as you can from within your company, you may decide to do some research beyond the corporate walls. This can be as simple a task as reading other companies' documentation—especially manuals that you think your users may be familiar with—or as difficult as going out to meet the users in person.

The best thing you can do to find out about your audience is to meet them. An easy way to do that is to attend current customer training classes. These are usually given close to your office, conducted by training personnel with whom you want to talk in any event, and filled with the exact people you will be writing for. You may or may not be able to ask them questions. But you will be able to watch them work with your company's product, listen to their questions, their frustrations, and their raves about things they like. If your company has a user-testing laboratory, that is also a place where you can observe your audience actually using the product, and you can get a good sense of their education, interest, experience, and so forth.

Another way to meet the users is to go to their work places. Although this may not be possible because of financial or geographic constraints, a trip to the field yourself is the best possible way to learn about the people who will read your documentation. If you have this opportunity, take with you a checklist of things to look for and the audience worksheets. Talk to the users about how they like using your company's documentation, or even another company's documentation. Ask what they would like to see, but don't make promises. The answers you receive go a long way toward giving you a sense of effective design, style, size, and other qualities that make your manuals more usable.

Ask your supervisor about arranging these site visits. It's best to go with field support people rather than with sales people because you want to gather information and to let the customers know that their input is important. If you go on a sales call, you may wind up listening to the salesperson's pitch rather than the customer's feedback.

PERSONAL EXPERIENCE

And, finally, think over your own experience. Any experience you've had with technical documentation will give you some insight into how your audience might use the material you're writing, but it may also hinder your ability to see things from the users' perspective. Think carefully about your own biases and your own expectations about computer manuals. Are you a former programmer who enjoys reading a lot of technical information? You may be tempted to include more of this information than the user wants or needs. (If you are just learning the product, you may be tempted to turn a reference guide into a tutorial.) The same is true for other preferences you may have that stem from your own experience and may not be compatible with the users'. Analyze your own approach to documentation and try to determine how well you can put yourself in the users' shoes.

EXERCISES

Complete all four of these steps:

1. Fill in the audience worksheets for both your primary and secondary audiences.
2. Write a prose description of your primary audience. If you have met a member of your audience, use his or her name and think of that person as you write the description. Add insights and details that did not fit in the worksheet. You should put on paper everything you know about that audience.
3. Distill this portrait into a one-paragraph description. You are abstracting from your longer description and writing the *formal audience description that will appear in your Documentation Plan.* Unless your manual is ultimately going to be used by only a few people, there is no need to use specific people's names in this more formal paragraph.
4. Photocopy this description and pin it to your bulletin board or tape it to your wall or computer. Throughout the rest of the writing process, keep in the front of your mind that you are writing to *this* audience. Such a visible reminder will keep you on track as nothing else will.

DEFINING OBJECTIVES

Once you have gathered as much information as you possibly can about your audience, you need to turn your attention to the document's objective. The question now changes from *"Who* will use the documentation?" to *"Why* will anyone use the documentation?" It's not enough to know everything about the user if you don't know what that user needs the document to do. Writers who have not thought about both audience and objectives often end up writing to themselves.

For example, one writer who was fascinated with the C++ programming language took a 250-page user guide, filled it with software information, and turned it into a 500-page "monster." This is programmer-to-programmer writing, and, while it works in some highly specific instances, in this case, the writer missed the point of a user guide.

To avoid making a similar mistake, make sure you spend enough time talking to people about what the documentation is supposed to do for the users. Think about the document's *objective* and state it in behavioral terms. By asking the question, "What tasks do the audience plan to *do* using the document?" you will come closer to determining the document's purpose. When you are talking to these resource people (your supervisor, the product developer, the marketing director, and so forth), listen carefully to the *verbs* they use to describe the document. Watch for key words such as "operate," "program," "repair," or "sell." Remember that there is not always a one-to-one correlation between audience and purpose; that is, just because you know your audience comprises systems operators and data-entry operators does not necessarily mean that you should write a manual that is solely general operator's instructions. Perhaps these users need the manual for other reasons as well. Make a list of the verbs they use as they answer your questions, and you will know more specifically the kind of documentation you need to design.

══COMMON OBJECTIVES FOR COMPUTER DOCUMENTATION

Chapter 1 provides a general overview of the kinds of computer documentation most writers are assigned to write. But these broad categories do not cover everything writers need to know about purpose. In fact, defining a document's objectives requires careful thought and a good deal of corporate "savvy," because most computer company's documents have dual purposes, not all of which are stated. Some of them are internal to a company and are fairly subtle; others are for the external audience and are more overt, often stated in the document's preface. Remember that every document contributes to the company's image. It is important for writers to realize at the outset the various goals their work is expected to achieve. Figure 2-3 illustrates the main objectives the document may serve for both the company and the audience, and these are also discussed in the next two sections.

COMPANY'S OBJECTIVES

A company chooses to document its products for many reasons:

1. *To inform customers.* Documents fitting this category define and describe a product to help customers make decisions about it, operate it, or maintain it. More often than not, a customer cannot use the product without online or printed documentation, or both. Hence, it is essential for the company to provide this kind of information.
2. *To market products.* A marketing representative who can show a prospective buyer handsome, easy-to-use documentation may have

---

TWO SETS OF OBJECTIVES

COMPANY'S OBJECTIVES

| To inform customers | To market products | To enhance corporate image | To state legal limitations for product | To save company money |

AUDIENCE'S OBJECTIVES

| To make a decision about a product | To operate or maintain a product | To program or analyze a product |

■ **FIGURE 2-3**   Objectives of Computer Documents

an easier job making the sale or lease agreement than the representa-tive selling either hardware or software with intimidating, hard-to-use instructions.

3. *To enhance the corporate image.* The style and appearance of online and hardcopy documentation perpetually does marketing duty—if they are good. For example, the Apple logo is instantly recognizable and is omnipresent on all Apple computer documents. The documentation itself has a distinct character all its own: it is light, friendly, and easy to use.

4. *To state legal limitations.* Documentation is also a place where compa-nies can state the limits of their legal liabilities and rights for a prod-uct. On the first page of most manuals appears the legal statements concerning proprietary rights, corporate liability for damages, and any other legal disclaimers.

5. *To save the company money.* As the price of computer hardware and software decreases, a manufacturer can no longer afford to send out trained representatives to help new customers with set-up, installa-tion, operation, and troubleshooting. Training has also become increasingly expensive, both for companies and customers. Consequently, companies are relying on well-designed and well-writ-ten documentation to take the place of expensive "house calls."

Although you will never see a document that says, "We printed this manual to save ourselves money, give ourselves a classy corporate image, and make you want to buy more of our products," these purposes are inherent in computer documentation and writers should be aware of them.

AUDIENCE'S OBJECTIVES

These are the overt objectives stated in the beginning of most computer manuals, or clearly expressed by the very nature of the documents themselves. It is important for the users to recognize these purposes so they can choose the correct documents for the jobs they want to do. Because these categories are so important, they require more detailed explanation.

1. *To make a decision about a product.* People who need documents of this type are often making a purchasing decision, but they may also be planning a system's use with a company or doing other tasks that require decision making. As a result, documents in this category have a broad range: from marketing/public relations material such as annual reports, news releases, and product announcements to highly technical information such as technical specifications, system introductions, and concept books. Often, these documents are designed to be read from start to finish. The more technically complex manuals are usually used for both reference and planning, and are therefore designed to be used for initial reading and future reference.

2. *To operate or maintain a product.* The audiences for these documents actually use the computer hardware or software. Because they are performing tasks in varying degrees of sophistication, they need step-by-step instructions for the necessary computer procedures. Documents in this category range from training material such as videos, online help, and hardcopy training manuals to complex systems and hardware manuals and simple user guides for home-computer users. These documents may be both online and hardcopy, and are rarely read from start to finish, except in training programs. In many cases, these instructions may double as user and reference guides, or they may be used once and filed away, as in hardware set-up instructions.

3. *To program or analyze a product.* This category includes documents that assist programmers, systems analysts, and sometimes engineers who program computer systems. These users are usually sophisticated in education and in computer experience, and they will use the documents repeatedly for reference but will never read them cover-to-cover, front-to-back. Included here are software systems manuals, language manuals, software utilities manuals, and applications software manuals. Other special documents, such as conversion manuals that give information of moving from one system to another, also belong in this category. Generally, these documents are compendiums of information and are too bulky to reside entirely online, although error messages and help texts may be electronic. Like encyclopedias and dictionaries, this documentation is usually organized alphabetically or according to logical sequences of functions.

To double-check your understanding of your book's purposes, fill out the Document Objectives Worksheet provided in Figure 2-4. Begin by marking all of the purposes you think apply to your audience. By looking at the purposes you have checked on the worksheet, you will be able to determine better the main goals for the document you are going to write.

===== EXERCISES =======================================

1. Fill out the Document Objectives Worksheet (Figure 2-4) for the document you are writing. Remember to identify the objectives of both primary and secondary audiences. Based on the completed worksheet, determine the main objective of your document and write it down in two or three sentences. *You will include this statement in both your document plan and in the document itself.*

2. Determine all prerequisite experience, necessary hardware and software manuals, and other tools your audience will need to use your documentation. Write these prerequisites in a sentence or two and add them to your statement of purpose.

3. Find a computer manual that has a stated purpose in the front of the book and evaluate how successfully it accomplishes these goals. Explain your analysis thoroughly. Next, try to determine the company's purposes for publishing the document. Does the manual accomplish these more subtle goals? Again, explain your analysis in detail.

FINAL THOUGHTS

Understanding audience and the document's objectives allows you, as the writer, to have all the preliminary tools you need to begin planning the document. Once you know to whom you are writing and what they will use the documentation for, you can develop a focused plan that aims directly at the user's needs, and you can speak with authority about what you should include and what you should exclude to make the book most useful for the people who will use it. Without this information, writers waste a lot of valuable time designing aimless manuals that will have to be revised extensively before they can provide genuine help to the user.

DOCUMENT OBJECTIVES WORKSHEET

Name of Document:	Primary Audience:	Secondary Audience:

Audience Objective: To make a decision about a product
Document should:
- Inform
- Aid in understanding
- Define
- Describe
- Present concepts
- Aid in making purchase decisions
- Aid in planning for systems use
- Other?

Audience Objective: To operate or maintain a product
Document should give instructions for:
- Daily operating
- Setting-up and installing
- Troubleshooting
- Testing
- Converting
- Planning
- Other?

Audience Objective: To program or analyze a product
Document should give information on:
- Preparing and updating programs
- Understanding language elements and rules for combining them
- Compiling, listing, executing, and maintaining programs
- Understanding operating systems
- Analyzing systems (such as database structures, network topology, etc.)
- Other?

FIGURE 2-4   Document Objectives Worksheet

CHAPTER 3

CONDUCTING A NEEDS ASSESSMENT

Sometimes technical writers are faced with a situation so complex that determining the audience and purpose for a specific document becomes secondary to clarifying the larger role of the documentation as a whole. In such instances, writers may be called upon to conduct a needs assessment. This task is similar to defining the purpose of a single manual, but it is done on a much larger scale and involves asking more global questions. Because of its comprehensive nature, a needs assessment is usually done by a project manager, but all writers should know how to proceed with the job. In some small companies or in freelance writing situations, one writer may be the only person available both to determine the documentation needs and then to write the manuals.

===== WHAT IS A "NEEDS ASSESSMENT"?

Basically, a documentation needs assessment is a carefully structured analysis of what the company's documentation must include to satisfy the requirements of both the company and the user. By systematically gathering information from the appropriate sources, you can provide a clearer picture of what that documentation should be, and you can make recommendations based on specific evidence.

The key to this procedure is the word "systematic." It's all too easy to take a quick look at the existing documentation, talk to a few people in the technical publications department or to a few developers, and then form hasty conclusions. That slapdash assessment may save time, but it will not give you enough information to make a valid and complete assessment of the situation. If a physician diagnosed her patients' medical problems in this haphazard way, would you trust her subsequent medical advice? Probably not. As a technical communication specialist, you should be every bit as careful as a good physician when you diagnose the communication problem that faces you.

As the preceding analogy suggests, a needs assessment stems from the presence of a problem to be solved. Consequently, the first step is to articulate the problem. For example, your company is planning to release its second software product. The documentation of the first product was so expensive—and of such limited usefulness—that the company will not continue to develop documentation like the previous model. The vice president of your division has asked you to reduce the cost of the new documentation by 25 to 50 percent and also improve its quality. Such a scenario is all too frequent in today's difficult economic times.

You know you can't address this problem without more information. With the vice president's mandate clear in your mind, you return to your desk to design a needs assessment.

═══PROCEDURE

There are many different ways to analyze needs, but some basic steps are common to the various assessment tools used:

- Identify the assessment's purpose
- Determine the assessment's limitations
- Select the assessment's target population
- Develop a questionnaire
- Field test the questionnaire
- Schedule interviews
- Collect data
- Analyze data
- Request further input
- Make recommendations

1. Identify the Assessment's Purpose

In this first step, you should think about the specific goals of the assessment. What exactly do you want to find out? Define the scope of the project as narrowly as you can while still soliciting as much information as you need to make trustworthy recommendations. Be careful not to phrase your purpose in a way that presupposes the results. In other words, try not to say "My purpose is to discover why the applications sections of the company's manuals are weak." Such phrasing already assumes that the manuals are weak, and an assessment built on that unsupported premise will produce equally tenuous results. Instead, state the assessment's goal in unbiased language that contains no hidden assumptions.

EXAMPLE: *Purpose*

To determine the strengths weaknesses of the existing documentation and to assess which features are most expensive.

2. Determine the Necessary Limitations of the Needs Assessment

Think about the things that limit your ability to conduct the assessment. How much time do you have? Do you have enough money to survey a large target population? Do you have enough other resources available to allow you to do a full-scale analysis? Once you have made a list of what you have to work with, you can design the needs assessment to best match your resources.

EXAMPLE: *Limitations*

Time: 30 days
Money: No additional monies allotted
Other Resources: Unlimited photocopying, one computer (with graphics software) for creating the questionnaire and final report, electronic mail service, limited time in people's schedules, no secretarial help, limited access to end users.

3. Identify the Target Population(s) for the Assessment

If you have focused your purpose narrowly enough, you can request information from a fairly well-defined group of people. On the other hand, if your purpose is broadly stated, the range of people you need to question will be broad as well. For the sample problem used here, the target population includes two segments.

EXAMPLE: *Target Populations*
- People who use the documentation (end users, users within the company, people within the company who deal with end users) are the best people to answer the question of the *usefulness* of the previous documentation.
- People who deal with budgets and finances (accountants, purchasing agents, vendors) are the best people to answer the question of the *costs* of the previous documentation.

4. Develop a Questionnaire

Questionnaire design is a topic too large to be covered fully in this book. Nonetheless, the several key points covered here can help your survey deliver more effective results. First, word the questions clearly so that all the respondents will interpret them the same way. Second, ask for only one piece of information in each question. Third, make sure the wording of each question does not imply an answer—you want to encourage unbiased responses. Fourth, include a range of response options that are mutually exclusive enough to cover every conceivable answer. Fifth, design the questions so that the answers will be easy to tabulate. (Avoid "why" questions because they lead to speculation rather than fact.) And finally, if you want to include "open-ended" questions that require the respondent to write opinions, put these at the end of the survey.

WEAK EXAMPLE: Is it desirable to include conversion tables in the manual?
                _____Yes   _____No

BETTER: To what extent do you use the conversion tables in this manual? Circle the number under the best answer:

Never	*Sometimes*	*Usually*	*Always*
1	2	3	4

Do you find these conversion tables helpful?

Never	*Sometimes*	*Usually*	*Always*
1	2	3	4

Note that this example includes *four* possible responses, thereby eliminating the possibility of a noncommittal middle answer.

5. Field Test the Questionnaire

A preliminary test of the questionnaire you have designed will reveal trouble spots you can edit prior to sending out the official survey. Field testing involves sending the questionnaire to a sample of people as similar as possible to the target population. Have as many people as you can critique and proofread the questionnaire, and encourage them to make comments and suggestions on all aspects of the survey. In addition to gaining valuable information from the test audience's critiques, you can also learn a lot by analyzing the results of the field test just as you would the final results. By doing so, you can identify items that yield unusable information and can eliminate them from the final questionnaire.

6. Plan an Interview Schedule

In a busy corporate environment, you will have limited time to conduct a survey. Moreover, written information is difficult to collect because of the strain it places on other people's time. One way of getting the information you need is to set up a schedule of interviews with members of the target population. Keep the times short, and come prepared with specific questions. You may use the interview to ask the questions listed on the questionnaire, or you may decide to proceed more informally and just listen to the person's thoughts about the situation you are assessing. It's a good idea to publish the interview schedule so that everyone involved knows the other players and realizes the importance of the task.

7. Collect Your Data

Now that you have a well-designed questionnaire and/or a published interview schedule, you are ready to actually collect the information. It is not imperative that you sample the entire target population. You can use a random or a stratified sample with equally effective results. Additionally, make sure you check all *written sources* that can offer valuable information (letters from users, product reviews in magazines, market surveys, user-response cards, and so forth).

> EXAMPLE: Choose to collect information from some—not all—salespeople, trainers, quality assurance personnel, hotline support personnel, and marketing people. Devise a method of randomly choosing the people you will question, rather than selecting them based on their personalities or proximity to your office. Stratify the sample by including respondents from all levels of each department you survey.

8. Analyze the Resulting Data

Analyze your data in such a way that it lends itself to feedback and sharing. Try using easy-to-understand formats such as tables or charts. When possible, report your findings in percentages.

EXAMPLE:

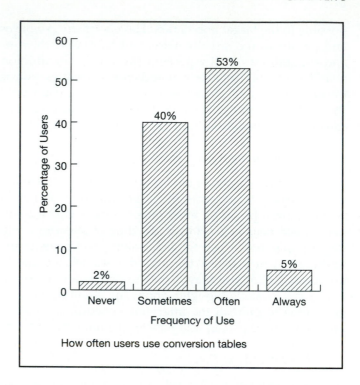

How often users use conversion tables

9. Request Further Input

After you have analyzed the data, send a memo to all who participated in the needs assessment telling them what your findings were and thanking them for their participation. Make sure you explain once again what the goal of your assessment is so that the results are placed in a meaningful context. Include graphics (tables, charts, and so forth), and invite rebuttals or further information.

10. Make Recommendations Based on Your Findings

If your needs assessment has been successful, you should have enough information to draw some significant conclusions. At this stage, make sure you do not slant the evidence to support your own biases, but do offer specific reasons for the recommendations you make. Attach copies of the questionnaire and tallied results.

EXAMPLE: TO: Chris Smith
                  Vice President, Software Development

            FR: Lee Curry
                  Manager, Technical Publications

            RE: Recommendations for Release 2.0 Documentation

Based on a carefully conducted needs assessment, I recommend that the company:

- Eliminate the hardcopy tutorial manual for Release 2.0.
- Embed more online help in the software.
- Redesign the user guide to include more quick reference material and less theory.
- Reduce the size of the user guide from the large, spiral-bound book to a 6 x 10 three-ring binder format not to exceed 200 pages.
- Eliminate glossy photographs and color.
- Include more examples.

Taking these steps will increase the usefulness of the documentation while simultaneously saving the company approximately 30 percent in documentation development and production costs. A complete discussion of each of these points and the rationale behind them follows. The results of the needs assessment appear in Appendix A.

FINAL THOUGHTS

Understanding how to conduct a needs assessment in a professional manner is an important skill for technical communications specialists for a variety of reasons. Not only does it allow writers to make trustworthy recommendations in difficult situations, such as in the case given above, but it also gives writers a tool they can use to assess the documentation's effectiveness at any time. Periodic needs assessments are a means to productively evaluate how well the documentation is meeting the needs of the company and the users. Further, conducting such analyses and keeping them on file creates a product documentation history that is extremely useful for future writers and corporate decision makers.

EXERCISE

Following the steps suggested in this chapter, conduct a needs assessment for a project you are working on. Your first action should be to define the goals for the assessment and write them down carefully. Then, proceed through the various stages, designing as large or as small an assessment as is appropriate for your resources. Make sure the final package includes a statement of the problem, your recommendations, the complete evidence to support those recommendations, and the information- gathering tools you used (questionnaires, interview questions, interview schedules, follow-up input, and so on).

CHAPTER 4

DEVELOPING
A DOCUMENTATION PLAN

After the preliminary research on audience and purpose is completed, the next step is to write a documentation plan: an outline that includes specific plans for the chapters as well as information on deadlines, production costs, and necessary resources. At this stage, many writers become impatient and want to get on with the business of writing rather than spending more time with preliminary work. But developing a good documentation plan will save time in the long run and will also make the entire writing project a more integral part of product development.

════WHY DEVELOP A PLAN?

There are two reasons why developing a documentation plan is a good idea. First, the outline helps you organize your thoughts before you begin to write, thus saving you significant revision time; and second, the plan helps all members of the project team understand and anticipate your needs.

As in any kind of professional writing, taking the time to plan ahead before plunging into the document itself allows writers to collect their thoughts and arrange them into the most logical sequence. Without an out-line—even a jotted one—writers tend to meander for a few pages before coming to their main points, and often the sequence of these points does not suit the needs of their readers. When writing for the computer industry, you should be especially careful to write with the user in mind, and that means planning in advance what you will say and how you will say it. The false assumption that you can sit down and write a perfect set of instruc-tions the first time through usually results in confused users, substantial time lost on editing, or both. The time spent on planning up front is key to producing the document quickly and efficiently.

If you are writing in a large corporate environment—or even in a com-pany employing just a few people—developing a documentation plan and distributing it to all interested parties clarifies your role in the project and makes your work easier. In many companies, a writer assigned a new docu-ment or an extensive revision of an existing document *must* submit a docu-ment plan for approval. This plan alerts all related departments and indi-viduals to the nature of the project and its time and budgetary requirements. The "doc plan" is also a contract the writer makes with the company: It describes the product to be delivered and the time frame for delivery.

On the other hand, writers in some companies—especially small ones—are not expected to produce doc plans. Their entire writing time may be as short as one or two months, and the company argues that there sim-ply is not enough time for such preliminaries. Nonetheless, as soon as the writers get started on the project, they realize that theirs is a seemingly invisible task—no one knows they need time on special equipment, or that they need product specifications, or that they need to be involved in devel-opment review meetings. No one at all knows that they need graphics for the manual, or that production and printing schedules are an issue in pro-ducing the document for the customer. As a result, the project takes much longer to complete than originally intended.

A documentation plan makes the writer's tasks visible to the rest of the company. By making the writing task a matter of public record, the doc plan helps integrate publications development with product development and creates a unified approach to all aspects of the project. Depending on time constraints, you can write the plan in two pages, or you can generate a more comprehensive plan that includes significant detail. Some large com-panies (Digital Equipment Corporation, for example) have doc plan "tem-plates" that all writers follow to assure uniformity in corporate documenta-tion. In small companies, the plan may be a memorandum that looks quite informal but still contains the vital information so that everyone involved with the project is alerted in advance to your needs and schedules. Make sure that everyone involved with the project gets a copy of the plan.

WHO SHOULD BE INVOLVED IN THE PLANNING STAGES?

A frequent question writers have as they begin to plan their documents is who should participate at this stage. Although you want to distribute the finished plan to all members of the project team, should all of them sit down together and help you plan? Not necessarily. Too many "cooks" at the beginning can definitely create confusion, especially if they are all in the kitchen at the same time. Instead of asking everyone to physically sit with you and construct an outline, it's a better idea to send a memo to all con-

---

DOCUMENTATION PLAN APPROVAL FORM

Document: *HMO REFERRAL DATA ENTRY TUTORIAL*

Principal Writer: *PAT BUTLER*

Product: *EDK HMO REFERRALS SOFTWARE*

Please initial this form and return it with your comments by **6/25/93**. *Initialing this form means you are responsible for providing the necessary information and meeting the required deadlines listed in this plan.*

Technical Accuracy   Approval
Jesse Smith, Senior Developer   *JS*
Lee Caton, Product Engineer   *LC*

Editors
Paula Brown, Senior Editor   *PB*
Terry Kovitch, Project Editor   *TAK*

Graphics
Del Simpson, Graphic Designer   *DS*

Printing
Cory Hesel, Production Coordinator   *CH*
Jack Corbett, Printing Services   *JC*

Customer Training
Chris Walsh, Product Trainer   *CW*

FIGURE 4-1 Documentation Plan Approval Form

cerned asking them to send you any preliminary input they may have on the project. Suggest in the memo that you are in the planning stages and that you will have a complete outline to them soon, but you would appreciate knowing about any potential scheduling difficulties they can foresee or other things you need to be aware of as you begin the process. If you are revising an existing document, this memo offers a good opportunity to ask people what they want to see changed in it.

Then, after you have received this information, sit down with a small group—preferably including an editor, a graphic designer, and perhaps a technical advisor—and put together the documentation plan. This task may take an afternoon, or it may take several days, depending on the size of the plan you propose, but it is extremely helpful to include these other perspectives at the beginning so that these people can continue to be helpful to you throughout the writing process.

One technique that may be useful at this stage is the "storyboarding" procedure. This technique is described in detail at the end of this chapter and can be incorporated at these preliminary meetings as well as at later meetings during the actual writing process. Please read the section on storyboarding before you complete your own doc plan. It may be helpful.

It is also important at this early stage to design a cover sheet you can attach to the final draft of the plan. This "sign off" sheet attached to the front (see Figure 4- 1) gives you a record of everyone who has approved the plan. Placing their signatures on the sheet encourages people to feel more of a vested interest in the project, and their recorded approval may serve to resolve conflicts that arise later in the documentation process.

═══ EXERCISES ═══════════════════════════════

1. Make a list of all the people who will be involved in your project. These are the people who will be on the "distribution list" for your documentation plan. If you are in a classroom situation, make a list of people you will need to use as resources for your writing project. Note specifically what kind of input you need from each. An editor, for example, can advise on department standards and organization as well as on grammar and punctuation. An engineer may have a specific expertise in a given area that you need to tap. Later, you can write an individualized cover letter to each person, focusing that person's attention on a particular area.

2. Design a Documentation Plan Approval Form you can use later as sign-off sheet for your doc plan.

▬▬COMPONENTS OF THE DOCUMENTATION PLAN

The documentation plan should serve as a road map for writers working on the document and should include as much information as possible about the text itself and about the other constraints affecting the writing process. For example, an effective doc plan may include the following parts:

- Identification of the document.
- Table of contents and overview.
- Audience definition.
- Document outline.
- Relationship to other documentation.
- Production information (format, graphic requirements, and so forth).
- Resource personnel.
- Your needs within the company.
- Schedules and milestones.
- Budgets.
- Potential problems and considerations.

IDENTIFICATION OF THE DOCUMENT

This first segment of the documentation plan includes the document's title, the order number (if there is one), and the product identification (for example, "IBM System/38, Release 3.0"). If the manual is a revision of a previous release, make sure to include a revision history so you can track any changes made in either the document or the product, thereby allowing you to understand better what additions or deletions you need to make. This section is also the appropriate place to put your name along with the other members of your writing team.

TABLE OF CONTENTS AND OVERVIEW

The Table of Contents (TOC) means just what it says. You should title and list the chapters you will write, including any appendices and glossaries. Immediately after the TOC, write a brief overview of the document. This is a prose description of what purpose it will serve for the company, what the document will contain, and what needs it will meet for the user. Placing such a description in the doc plan gives the reviewers a clearer picture of your project so they can contribute more helpful suggestions as they read the plan.

AUDIENCE DEFINITION

It is important to know who your audience is. Including this information in the doc plan justifies your decisions about your manual's design and contents. Again, you want to state this specifically so the reviewers will not have to guess about your readership and can review your plan more effectively. If you have completed a thorough audience definition as described in Chapter 2, you can incorporate a shortened version of what you wrote then into this section.

DOCUMENT OUTLINE

While this section may sound similar to the Table of Contents, it is more of a prose discussion of the various components of the document. Explaining your goals for each chapter allows both you and the reviewers to make sure you are planning to meet the users' needs with your manual. Note that the outline can be in whatever form best suits the document, the reviewers, and the author. It need not be formally structured with roman numerals and short, cryptic entries. Sentence outlines, paragraphs followed by short entries, or combinations of these techniques are acceptable, as long as the outline itself gives reviewers a good sense of the document's content and organization. Figure 4-2 shows a sample of such an outline that gives reviewers a clear idea of the writer's plans for each section.

RELATIONSHIP TO OTHER DOCUMENTATION

Understanding how your manual relates to others within the company as well as to external documents is important to both large and small companies, though probably more critical to the large ones. A small start-up company may never have produced other documentation, so there are no other manuals with which to compare this document. However, in such an instance you may research other companies' documentation and plan to make your manual complementary to it, allowing the users to feel comfortable making the transition from one document to another.

Within a large company, a manual may be a part of a family of documents known as a "documentation set." Noting this fact in your doc plan lets the reviewers and managers know that you and other writers have carefully considered how the pieces of the doc set relate to one another without duplicating or omitting critical information. For example, you should include the rationale for planning both a tutorial and a user's guide, or a quick reference along with a reference manual. (Usually, the documentation project manager will have written a large doc plan for the entire set—check to make sure your individual plan matches the overall design.) At issue for the reviewers will be *cost*—in terms of duplicate pages, printing, and binding—and *customer confusion* if the same information is printed in two or three documents. Provide clear reasons for your decisions.

Reviewers also may want to know the relationship of your document to the competition's. It is important to know who your competitors are and

---

HMO Referrals Data Entry Tutorial

Preliminary Outline

The following is a preliminary outline of the *HMO Referrals Data Entry Tutorial*:

1. Introduction

    This first part of the tutorial explains to data entry operators what an HMO system is and what function it serves. In addition, users learn what a referral is and how it functions within an HMO system.

2. About This Tutorial

    Part Two introduces users to the tutorial. This part explains: who should use this tutorial, what users will learn, and a brief description of the tutorial's lessons' content and objectives.

3. Lesson One: Getting Started

    The third part of the tutorial begins with Lesson One. The lesson's objective is to get users working with the referral function right away. It first explains important keys users need to understand before proceeding into the tutorial. Users then learn how to enter the referral function and create a sample referral by responding to referral function prompts. The final section is a summary highlighting major features.

    This first lesson contains sections explaining how to:

    *   get into the referrals function
    *   create a referral
    *   enter a sample referral, responding correctly to function prompts
    *   exit the referral function

    Lesson Two: Understanding Referrals

    In Lesson Two, users are given a more indepth view of the referral function. Users learn how to access sample patient accounts and create sample referral numbers for patient look-up. This lesson also provides users with a brief explanation of each referral prompt. Users create a patient account by entering information located on a sample referral form.

FIGURE 4-2   Preliminary Outline

how your document will fare on the market. Is your manual more user-friendly than others? Does it include more online information? Are there more features? Be sure to anticipate all of the marketing questions the reviewers may have and provide appropriate answers.

PRODUCTION INFORMATION

The amount of production information you include will vary according to the company and according to how much of the production you will have to do as a writer. In large companies with production departments, graphics departments, in-house printing or access to printing vendors, and shipping departments, you may need to specify only the type of document,

---

HMO Referrals Data Entry Tutorial

The final section of Lesson One is a summary outlining key features of the lesson and a one page step-by-step procedure for entering a referral.

Lesson Two contains information for a user to:

- access patients using the "lookup" method
- generate a valid referral number
- respond correctly to referral function prompts
- use the online help key ("?") for additional information
- enter a referral from information located on a referral form

Lesson Three: Editing Referrals

In this final lesson, users learn how to make changes to referrals using system editing features. Users first enter a sample referral from patient information located on a sample referral form, then review the referral to make changes to it. The lesson then explains how to respond correctly to system error messages, access a referral, and call up the system's information options menu, and choose a referral option to work in. Also provided is a section describing how to delete referral questionnaire prompts and patient records. Like the other lessons, the final section of Lesson Three is a summary of important procedures discussed throughout the lesson and a one page step-by-step procedure for editing a referral.

This last lesson contains sections explaining how to:

- enter a sample referral
- get into a referral review and change system responses
- call up the system's information options menu
- access a referral with a claims activity posted against it
- delete answered referral questionnaire prompts and records

Glossary

This final part of the tutorial/manual is a section containing definitions of system features and referral prompts used throughout the document.

---

approximate number of pages, desired paper stock, required graphics, and any other special printing requirements. Production will handle the rest.

But in a small company, you may be the production department as well as the writing department. In this case, you must supply more detailed information in the doc plan, and you may need to work with your editor and with outside vendors to get the complete picture. The necessary information includes:

- Approximate number of pages.
- Type of production (that is, whether the document will be typeset, desktop-published, or produced by another method).
- Paper stock.

- Typeface or font.
- Use of color (one-, two-, three-, or four-color?).
- Type of binding (perfect-bound, spiral-bound or saddle-stitched?).
- Kinds of graphics (line drawings, schematics, screens, or charts?).
- Tabs.
- Pullouts.

RESOURCE PERSONNEL

List here all the people you need either to give you information about the product, test the documentation, or help you produce the document. For example, include here names of developers, production staff, customer services people, human factors and quality assurance testing personnel, editors, reviewers, and so on. You may have some special information sources (such as libraries, external consultants, beta-test-site personnel, and so forth) that require you to travel or to spend money to transport the information to you. Identify these sources here.

Remember to use *specific names* at this stage to alert people that they have a responsibility to help you and to encourage in them a sense of commitment to the document. If people's names are in the doc plan and they have signed off on it, they are more likely to honor your requests later when you ask them to give you information in a timely fashion. Remember, too, that a brief, polite, personal contact is more likely to elicit a helpful response, especially if you can assure the person how important his or her input is to the document.

YOUR NEEDS WITHIN THE COMPANY

Take some time to really think about what you will need to complete the document: Are product specifications, access to software, access to a prototype of the product (if it's a new product), access to computer time, or access to specific desktop publishing equipment among these needs? If you will need to work special hours in order to complete the project, record those hours in this section.

Here, too, you should identify any special meetings you need to call or attend to gain information: review meetings, marketing strategy planning meetings, developers' meetings, and so on. You must remind everyone that you are part of the project team and should be invited to all relevant meetings about changes in the product.

Finally, if your company has other projects that are being developed simultaneously with yours that may affect your documentation (perhaps by competing for the same resources), you should have specifications on those projects. You need to know what is going on to make your document complete and accurate. For example, if you are documenting spreadsheet software while another group is documenting database software for the same product, it's a good idea for the two groups to be aware of each other's progress. And it enhances the company image and user "comfort" if they can agree on a common or similar approach and use consistent terminology.

SCHEDULES AND MILESTONES

Establishing the schedule is a key component in the doc plan. Before you commit to paper a timeframe for the document, make sure you have spoken to everyone involved with the project and have a realistic projection of the time needed to complete the manual. If you can, determine the ship date for the product and work backwards to schedule deadlines for the various components of the document. Be generous with your estimates, and try to schedule everything. Such thoroughness will commit other people to include your needs into their schedules and will alert you to potential problems with conflicts.

It helps to represent this schedule graphically in a timeline or in a chart indicating milestones for the project (See Figure 4-3). Such a visual representation is easier to read than is a prose description and it can be pulled from the doc plan or photocopied and taped up beside the participants' computers or desks.

Keep in mind that this schedule is effectively a promise to everyone involved that you will have the drafts and the other material to them by the dates you have suggested. They can schedule their time accordingly. And it is also a promise to the company that you will have the project completed by a specific time. It is therefore wise to be as thorough as possible in preparing this section of the documentation plan. In the interest of your own credibility, note what contingencies might affect the schedule—for example, "In order to meet this schedule, we must have functional specs by May 28 and a working prototype by July 1." Other people should know that you are depending on them. Likewise, if you see your own schedule slipping, inform everyone right away. Don't let a small slip go unannounced

FIGURE 4-3 Sample Documentation Schedule

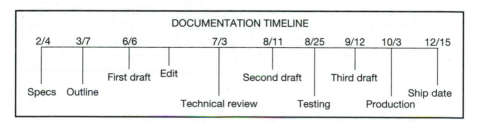

DOCUMENTATION TIMELINE

| 2/4 | 3/7 | 6/6 | | 7/3 | 8/11 | 8/25 | 9/12 | 10/3 | 12/15 |

First draft Edit   Second draft   Third draft

Specs   Outline   Ship date

Technical review   Testing   Production

and potentially grow into a major crisis. Notify people while the problem is still small and fixable.

BUDGETS

Some companies require preliminary budget estimates for each document, particularly if many writers are involved in a single project, or if graphics, typesetting, and printing are to be done by outside vendors. In larger companies, managers and project leaders do budgets; in smaller companies, writers may be asked for this information. Make sure to include all facets of production and any other cost items incorporated in the writing project.

POTENTIAL PROBLEMS AND CONSIDERATIONS

Effective documentation plans serve your needs and the needs of the company at large because they encourage the efficient flow of information and synchronize people's busy schedules. But even though they are so important, doc plans are not composed absolutely of the same parts in every company, nor are they static once they are drafted. No matter how carefully you plan for seemingly all contingencies, it is likely that the project will slip a deadline here or there, and the schedule will have to be revised. As frustrating as this is, be prepared for it to happen and remember to send notices of all slips to the project team members. If deadlines must be missed, at least everyone should be notified and asked to adjust their schedules. The goal, of course, is to have to send as few as possible of these "slip notes."

═══ EXERCISES ═══

1. Prepare a documentation plan for a project you are undertaking at work or in class. Try to keep the plan as brief as possible while including all the necessary information. Have the plan reviewed and approved or rejected by a group of at least two people before you proceed with the writing process. Use the Documentation Approval Form you designed for this purpose.

2. Visit two or three printers and obtain estimates for two different documents: (1) a three-fold brochure on good-quality stock, and (2) a 100-page manual with ten line drawings. Get estimates for both typesetting and photocopying, and include an estimate of how long it will take the printers to complete the job. Note how adding various bindings, color, tabs, and other design elements affect the price of the job.

STORYBOARDING

As mentioned earlier in this chapter, *storyboarding* is a technique that can help writers to collaborate in planning the organization of their documents. According to some technical communications experts:

> Successful documentation projects are team efforts. Writers need information from system developers and users, marketing and product staff, and technical experts. Getting their cooperation, feedback, and enthusiasm for a project is a constant challenge.... Storyboards are a valuable tool in the documentation development process. Written when a project is in its early stages, they can make a document *real* to users, reviewers, and managers. Reviewed and revised before a draft is written, they encourage participation, discussion, and reviewer involvement.
>
> Paula Berger and Martha Bednarz, "Storyboarding: Make It Real and Make Things Happen," ITCC *Proceedings*, 1987.

To create effective storyboards, writers must first have completed all the preliminaries of determining audience, purpose, and basic content of the document. They also must write an outline of the manual's contents, breaking out the topics into similarly structured components. These can be chapters, or subsections within chapters, or chapter groups, or whatever reflects balanced components of the document. Once that is finished, the actual storyboarding process begins.

The writers create a separate "board" (posterboard, flip chart, erasable blackboard or whiteboard) for each item in the outline. Each of these topics is presented in exactly the same format, producing a miniature model of the document compartmentalized onto a series of boards. The components of each are as follows:

- A *heading* that clearly indicates the topic.
- A brief *summary* of the component's text, written in the style and tone of the final document.
- *Notes* about the topic (questions, suggestions, examples, and so on).
- Suggested *visual aids*.

When this task is completed, the writers can invite the entire project team (including developers, graphic designers, marketing specialists, and so forth) to participate in a preliminary review meeting of the document. The group posts the boards around the room and can see how the whole project fits together. If the organization seems questionable, the team can simply rearrange the boards, creating a new organization on the spot. Then they can begin to discuss each board separately. Does it contain accurate and sufficient information? If not, the team can insert it. Is the style appropriate? If not, change it. Do the suggested examples make sense? If not, replace them with new ones. And so on.

This procedure allows the planning to be a group process that permits dynamic changes to happen immediately as a result of team input. Its strength is that it encourages collaboration and prevents problems caused when reviewers get their first look at the document during later stages in the writing process. Everyone involved in the project becomes familiar with it from the beginning and can start immediately to plan their own contributions. Later reviews are easier, and the writers can proceed with more confidence because they know they have a clear mandate from the entire team. On the other hand, storyboarding's weakness is that it requires a great deal of commitment and effort at the front end of the project. It also requires a group that can work together productively. If there are too many people in on the discussions, or if the conversations remain unfocused, the process may veer out of control. But handled well, storyboarding is an especially effective technique for document planning.

═══ EXERCISE ═══════════════════════════════════════

As you are completing your document plan, take the time to develop a structured outline that presents topics in similar formats. Then, following the steps suggested in this chapter, create a storyboard for your manual. Have members of your writing group participate in the discussion process, and actually incorporate their suggestions on the boards you have posted around the room. After you have finished the process, jot down what you think worked about it and what did not. Try to determine why certain things worked better than others. What suggestions do you have for changing the procedure next time?

CHAPTER 5

GETTING INFORMATION

When technical documentation managers and technical writers are asked to identify the biggest problem they have on the job, they respond unanimously: "Getting information from developers and engineers." Covert looks of understanding pass from face to face, and most writers sigh audibly. In both small and large companies, the difficulty of getting technical information about the product makes it harder for writers to do their jobs.

There are a number of reasons for this industry-wide problem, but the primary one is that the product often *does not exist* during the time the writer is supposed to be documenting it. It may be only a gleam in the developer's eye or an elusive note on product specs. And if the code hasn't been written or the prototype built, the developers really do have a hard time providing concrete information to the writers.

In the best situations, writers have access to prototypes of the product and development specifications, or perhaps a previous version of the product, and can learn about it from hands-on experience. But even that is not enough. Development specifications are often written by development engineers and are nearly incomprehensible to anyone else. And, the prototypes may be filled with "bugs" (technological problems) that prevent writers from accurately learning how the product works. No matter how much product information writers have written down in front of them, they will have many more questions than this material can answer. In short, they need to find ways to get information, or they cannot do their jobs.

SOURCES OF INFORMATION

There are a number of information sources you should be aware of before you begin writing, and a number of sources you can tap after you have written a first draft. It's important to use them in the order suggested here because the elements in the first list are primary and should be your first steps, whereas the elements in the second list may be helpful only after you have thoroughly done the primary research.

The preliminary sources are:

- Product objectives statement
- Product specifications
- Development review meetings
- The product itself
- Interviews
- Existing documentation
- Libraries

Other sources that emerge after your first draft are:

- Testers
- Field support personnel
- Editors
- Technical review comments

PRODUCT OBJECTIVES STATEMENT

The product objectives statement is frequently the first document produced about a product. It typically has a marketing orientation, in that it identifies the audience (potential market), the main functions of the product and the needs it serves, an approximation of how many units it may sell, what the competition is, and possible problems and contingencies. While not as technically detailed as the product spec, it does contain some information that is invaluable to the writer, and it is often available very early in the product development cycle.

PRODUCT SPECIFICATIONS

Specifications, or "specs," are the most standard source of information. In some highly structured companies, specs are written by developers and/or marketers as they conceive of a product. Before writers are assigned to do the documentation, they may be given a lengthy set of specs on everything from the movement of bits to error codes. But because of their immense detail, specs can often be 300 pages long, even for relatively easy-to-use products.

Another common problem with specs is that they may be hard to understand at first. They are written as records of what developers have been doing, not as customer documents—hence there is no premium placed on graceful, clear prose. Moreover, development programmers and engineers are paid to design products, not to write, and they may be good at explaining how to "build" a product, but not as effective at explaining how to use it. These problems combine to make the technical writer's task of translating this material difficult.

Nonetheless, the specs contain most, if not all, of the product data. With practice, writers generally find that they can understand the specs more easily than they thought at the outset, and the effort is certainly worthwhile. (You can make specs more understandable by enhancing your own technical background in the technology you're documenting through coursework, extensive reading, active listening, and hands-on experience.) The specs are a sort of product bible for the writer, and those writers who have access to them are much more fortunate than those whose companies do not produce specs at all. Figure 5-1 shows part of a typical product specification.

DEVELOPMENT REVIEW MEETINGS

Companies that have many developers, managers, testers, market planners, and other team members working on a product usually schedule project meetings. Some of these meetings may be informal, while others are highly structured. Regardless of how they are conducted, these meetings are essential for writers to attend; they are excellent sources of information.

These meetings provide writers with the opportunity to meet developers, project leaders, testers, and other significant people. They also provide the opportunity for writers to observe the dynamics between the members of the development team. This knowledge can be an important element in understanding who has the most accurate information, and it also reflects the politics of the group—something to which every writer should be sensitive.

FUNCTIONAL SPECIFICATION : 752

XYLOGICS INC.
144 Middlesex Turnpike
Burlington, MA 01803

Revision 1.0

1.   Product Description

This document describes the functionality of the XYLOGICS 752 VME/SMD-E
disk controller. The 752 controller, utilizing an 8kb FIFO buffer incorporates
multi-track read-ahead with intelligent DMA techniques (DYNA-THROTTLE) to
provide significant performance improvements over competitive products and
was specifically designed as a performance upgrade over the 751 controller.

Because of the similarity between the two products, this document will describe
the 752 in reference to the 751. The 751 functionality is provided in the XYLOG-
ICS MODEL 751 USERS MANUAL, Revision A.

2.   Software Interface

The 752 software interface is essentially the same as defined for the 751.
Commands are passed to the controller and executed in exactly the same manner as described in the 751 manual. The table below outlines the differences
between the 751 and 752 software interfaces. Refer to the following sections to
get a more detailed description of the differences between the 752 and 751 con-
trollers.

	751	752
• Controller Type Code	51	52
• Command Optimization Feature (COP)	YES	N/A
• Zero Latency Read (ZLR)	YES	N/A
• Disable Read Ahead Feature (DRA)	N/A	YES
• Write Format Parameters Required	NO	YES
• Read Ahead IOPB Status Bit (RAIOPB)	N/A	YES
• Interrupt End of Chain (IEC)	YES	N/A
• 450 Compatible Mode (C450)	YES	YES
• Dual Port Drive Support	YES	YES**

**NOTE: See Section 2.7 for restrictions.

FIGURE 5-1   Sample Page from a Product Specification (Used by permission of
Xylogics, Inc.)

If you attend these meetings and are knowledgeable about the product, you may contribute to the development process by offering valuable
insight to the discussions. If you are a new writer and know little about the

2.1 Controller Type

The 752 controller returns a "52" as the controller type (byte E) during a Read Controller Parameters command rather than the "51" returned by the 751.

2.2   Set Controller Parameters

The 752 controller does not support the COP and ZLR bits specified in the Set Controller Parameters command. The COP enables the IOPB reordering and elevator-seek features, which are not supported by the 752. The ZLR bit enables the zero-latency-read feature, which is not supported by the 752. Accidentally setting either of these bits will cause no problems as these bits are simply ignored by the 752 firmware.

2.3   Disable Read-Ahead Bit

The 752 controller features an inherent read-ahead scheme to improve read performance significantly. A new bit in the Set Controller Parameters command has been added to disable the read-ahead performed by the controller. Setting the DRA bit (bit zero (0) of byte 9) and executing the normal Set Controller Parameters command will cause all subsequent reads to be satisfied via disk data rather than the read-ahead buffer.

2.4   Volatile VS Nonvolatile IOPB RAM

The 752 controller requires a Write Format Parameters command be executed following power up and before any disk access can be performed. The 751 controller utilized a nonvolatile RAM that contained factory preset format parameters, which were generally used without modification by the customer. The 752 controller uses a traditional RAM part and must therefore be initialized with the correct format parameters following power up and before any disk access can be performed. An explanation of the exact format parameters to use can be found in the *751 USERS MANUAL*.

2.5   Read-Ahead IOPB Status Bit

The 752 controller may satisfy a read request via data in the read-ahead buffer. If a read IOPB is satisfied by data already in the FIFO rather than a traditional disk access, the IOPB will be updated with the RAIOPB bit set (bit 0) in the 752 internal status byte (byte 3). This bit has been added as a performance analysis tool to accurately measure the hit-rate of the read-ahead offered by the 752.

2.6   Fatal Error Codes

Due to the removal of the nonvolatile RAM on the 752, fatal error codes EO (IRAM Checksum Failure) and E2 (EPROM Checksum Failure) are no longer supported by the 752.

product, you will still be able to pick up some essential information just by being at the meeting and listening carefully.

Another reason for attending these meetings is that the developers will get used to seeing you, will come to understand that you have a role in the product's development, and consequently will be more likely to remember to give you information when you need it.

THE PRODUCT ITSELF

Above all, writers need time with the product they are documenting, whether it is a software package or a new machine. Unless you have used the product—have laid hands on the keyboard or cables—you will have a very hard time documenting it. And, if you are expected to write examples of how the software works for the customer, you simply cannot do your job without using the product.

How do you manage this if the product isn't complete? You work with the portion that is complete: the prototype machine, the finished sections of the software package, or the products closest in design to the one you are assigned. At this early stage, you may not be able write much, but you will be using the machine and/or the software itself, or seeing the screens and the commands that the customer will ultimately have to use. This hands-on experience gives you a sense of the final product and is immensely helpful to your understanding of your project.

Gaining access to the unfinished product may mean that you have to accommodate your schedule to that of the developers. Developers may work at night, and the only time you can see their "work in progress" is at 2:00 A.M. In the early days of the computer industry, it was common for engineers to work around the clock on new computer designs and new software, and the development labs were most productive after midnight. Those hours are less common now, although some start-up companies still work at feverish pitches to make a go of it in the highly competitive marketplace. As a technical writer for such a company, you may find you are working atypical hours to get your information or to get time on the system.

But such schedule contortions should be rare if you have written an effective documentation plan and have had it approved by all the people you need to see. Remember, the developers have signed off on your plan and should therefore be expected to provide you with the information you need.

You may also need time to work with competitors' products. Some companies purchase a competitor's computer and software to have it available for their staff. Gaining familiarity with these products allows you to better understand the expectations the users have when they shop around for computer products. Once you are aware of these expectations, you can design your documentation to be more competitive and to function in ways familiar to the user. You may also discover what sets your product apart from the pack and can emphasize those features as you write.

INTERVIEWS

Even when writers have access to the prototypes and a detailed set of specifications, they still need to talk with the people who know best how the product works: developers. It is much easier to ask questions of a person than of a pile of specs, and the answers may lead to other questions, which the developer can answer immediately. But developers are often reluctant to stop their work to talk, especially when writers do not have as sophisticated technical knowledge of the product as they do. Getting information from them can be a trial. Nonetheless, because consulting them is essential to the documentation process, an entire section of this chapter (pp. 58-59) is devoted to working with developers. They should be your primary source for technical information.

Other people can provide useful material: market planners, who may be working closely with development project leaders to design the project, and who themselves are working on documentation for marketing and sales personnel; testers, who must use the new hardware or software as soon as it is available to verify that it is a viable product; trainers and course developers, who know the problems people often have in learning how to use the product; and other writers, who may be writing companion documentation to your own or who have worked on similar projects in the past. After talking to the developers, you may want to interview some of these people to help you become even more familiar with the product.

As is true in all work environments, everyone's time is limited, and the more efficient use you make of it the more likely it is that your interviews will be productive. When you conduct interviews, be professional. Make appointments in advance if possible, explain what information you will need so they can prepare, and come prepared with a list of questions and an organized plan for the discussion. Afterwards, it's a good idea to write a brief summary of the information you received and send it to the interviewee for confirmation of accuracy. In any discussion, miscommunication can result, and it is important to take precautions against this happening. An extra benefit of this summary is that the resource person has a chance to add even more information—and you have it all in writing. One more tip: Don't forget the thank you. A little gratitude goes a long way toward cementing positive work relationships you can count on in the future.

EXISTING DOCUMENTATION

Another source of information is documentation produced by your company and by other companies—whether they manufacture competitive products or products compatible with your own.

Within your company, you may find documentation about the product you are working on or about related products. If you are documenting a new release or an updated version of a product, the existing documentation will provide you with a basis from which to start. These manuals—if they are good—will also give you a guideline for style, format, and audience level. If the documentation is not good, you may use the manuals as departure points or "bad examples" against which you can compare your version, always working to make yours more effective.

To some extent, you should check existing company documentation on other products. Companies often want a sense of continuity from one document to the next, even between different products, in order to create a recognizable corporate image. Make sure any changes you would like to make in the company documentation style are well supported with research on cost effectiveness and marketing appeal. There may be a corporate style guide that can provide some clues and/or shortcuts and answer questions about consistency. (See Chapter 9.)

Finally, you need to look at competitors' documentation. Understanding the competition allows you to make better decisions about your own project. You may want to emulate these documents or improve on their techniques, but in any case you should be as familiar with these manuals as your users will be.

LIBRARIES

Computer technology is developing faster than books can be written about it. As a result, you may think that a library may be of limited usefulness to technical writers. However, many libraries stock trade magazines and newspapers that contain the most current information about computer products—*MacWorld*, *PC/Computing*, *PC Week*, *Computerworld*, and many others are at the cutting edge of computer developments. Libraries also may provide access to CD-ROM, an incredibly powerful electronic research tool for finding out the latest information about a company, a product, or other essential data. You may use academic libraries to research technical communication journals and conference proceedings. If you are a novice at computer topics, you may also find books in the library to help you understand the theory behind some of the more traditional aspects of computers: artificial intelligence, programming languages, networking options, and so on. Such information will help you to be more comfortable with the technology, but be wary of including too much theory in your documentation— it may only clutter the operating instructions.

After you have used these technical sources to produce a first draft, at least four resources emerge to give you initial feedback on that draft: testers, field support personnel, editors, and technical review comments.

TESTERS

Testers may be from quality assurance, systems assurance, beta-test sites, and/or human factors, and they will probably begin working on a first- or second-draft document. They may be testing the manual for accuracy and ease of use, or they may be testing only the product itself. In either case, they have valuable information to add to the draft as they discover items that need to be changed both in the documentation and in the product.

FIELD SUPPORT PERSONNEL

People who work on site with the users can take drafts of your manual and test it with the audience you are writing for. Some field support people are the ultimate audience for your book and can give it a test run in the field before you produce your final version. Education and training personnel can perform much the same function if they are willing to use drafts of the book in classes or in on-site training sessions.

EDITORS

Editors should be involved in your document project from the initial development stage, so the first draft will not be entirely new to them. Nonetheless, they can certainly provide comments on the writing style and organization and should see the document at all of its stages. Chapter 10 discusses the role of editors in detail.

TECHNICAL REVIEW COMMENTS

Once you have finished a draft of your document, it usually circulates through a variety of reviewers from many departments in the company. Especially important are the comments you receive from the research and development department; they check the document for technical accuracy and can offer essential suggestions. Make sure they provide written comments you can refer to as you make revisions.

Note that you are more likely to get useful feedback from all of these sources if you include a short cover memo with your draft asking for concrete, constructive comments on specific areas.

=== EXERCISES ===

1. Make a list of the preliminary sources you have available to you. Then, list the sources you can use after you have completed a first draft. Be as exhaustive as you can, thinking of every possible resource for both lists. Beside each entry,

make a note about the kind of information that source will be most helpful in providing.

2. Make appointments to interview at least two people from your list of preliminary sources. Go to the interviews prepared with questions and take good notes. Afterwards, write a summary of the discussion and send it to the interviewees for confirmation. Don't forget the thank-you note. (Use electronic mail (E-mail) if you have access to it.)

3. Go to the library and research all of the sources there that may be helpful to you on your project. Check the card catalogues, the computer databases, and the periodical index. Make an annotated list of at least five sources, explaining briefly how each one is helpful.

4. Find documentation for a product that competes with the one you are writing about. Spend some time analyzing that documentation: What does it do well? What does it do poorly? How will yours be an improvement? What features will you emulate, if any?

CONSULTING DEVELOPERS

In the computer industry, the myths that have been perpetuated about development engineers are fantastical. The popular image is of wild-eyed technical visionaries, machine wizards, and workaholics who exist on Chinese take-out food, Classic Coke, and Twinkies, never seeing the light of day. They are given free reign to work whenever they choose and in whatever fashion they choose: in the middle of the night or on weekends, in t-shirts, blue jeans, and sandals and always accompanied by a stack of computer games. When they talk, they speak in tongues comprehensible only to the initiated. They aren't interested in anything but 1s and 0s. As a result, they often find talking to anyone below their level of technical understanding intrusive and counterproductive.

Nonsense. Certainly there are developers who fit this description, just as there are people who fit the stereotypes in any profession. Many developers have equally unflattering views of technical writers as unimaginative drones who have no technical knowledge and are not worth talking to. Unfortunately, the myths about both groups are so widely believed that working together can become much more difficult than it need be. The key to working well with developers—and the key to working well with "tech writers"—is to avoid wasting their time and to show genuine interest in their work.

Many technical writers are so intimidated or frustrated by developers that they avoid any direct contact with them. As difficult as it is to talk to developers, it's sometimes even more difficult to *find* them because they don't work at specific desks. Unwilling to make a concerted effort to locate them, some writers resort to electronic mail messages or memos and are angry when they get no response. Sometimes the only way to get information is to physically go in search of it. Writers who never leave the safety of their own cubicles are probably not doing their jobs very well.

To find developers and gain their trust, you have to be present at the development labs, at review meetings, at any place where the developers can begin to think of you as team member who is seriously concerned about the product. In some companies, writers are assigned to development teams and will actually move their desks to the development area to sit with the project engineers day in and day out. If you are in this situation, you're on the spot to participate in conversations and meetings about the product's development. After a time, you may begin to take part in the actual design of the product, contributing to writing the messages that appear on the screen, doing very early testing, and commenting on human factors elements such as usability and consistency. Thus you are both contributing to the product and documenting it simultaneously.

On the other hand, when writers sit together as a separate technical documentation department, they gain professional identification with other writers and benefit from the camaraderie and the support, although they must work harder to make contact with the developers. Constant phone calls and E-mail messages do nothing to encourage a productive relationship. No matter which approach your company chooses—a separate documentation department or integrated product teams—*you* are responsible for establishing open channels of communication.

When you finally track down a developer, be prepared to ask good questions. Many writers make the mistake of scheduling an appointment with the developer before doing any research or thinking about the product at all. This premature meeting usually yields nothing but inefficient conversation and bad feelings. Instead of rushing off to meet the developer right away, take some time to learn the product and to do any background study that might be useful. Then, when you do sit down with the developer, you can ask specific questions and understand the answers you receive. The developer will not have to waste time explaining things over and over again.

If you are lucky and indicate sufficient interest, you may find a developer who responds to that interest by giving you more information than you requested and by offering to read portions of the document before it is circulated for initial review. Such an offer is invaluable. Accept immediately and follow up on it.

WHERE NOT TO GO FOR INFORMATION

With all of these resources available, some writers may assume that everyone is fair game for questioning about the technical aspects of a product.

This is not true. Writers should realize that the best place to get technical information is to go to the primary source: the developers. Without their help, documenting a product would be nearly impossible. But because tracking them down can be time consuming, some writers may be tempted to take shortcuts: they may ask their managers, their editors, and other writers to do the legwork for them or to answer the technical questions as best they know how. In nearly every instance, asking these secondary resources is asking for trouble.

When you are assigned to write about a particular product, your supervisors assume that *you* are responsible for getting the necessary information. If your first step is to ask these supervisors—or ask other writers or editors—to supply technical material for you, you are imposing on their time and on their good will. Further, whatever technical knowledge they have is secondhand; it originally came from developers. Such information handed down from person to person can get less and less accurate the farther it gets from the original source. While you *should* talk to your project leader and other writers and editors on an ongoing basis, it is not a good idea to use them as a primary source of technical information.

The same caution applies to other members of the company who, in fact, may depend on you to get the information and pass it along to them in the form of documentation: sales representatives, customer support people, field service personnel, education and training personnel, and so forth. These people are not generally appropriate resources for technical information.

For what kinds of information can you rely on these very knowledgeable people? As Chapter 2 and Chapter 3 illustrate, they can provide insight into audience analysis, document purpose, format and layout, marketing strategies, style conventions, and writing strategies and may provide valuable input on drafts of the document. The trick is not to ignore these people as resources, but to use them in the right context.

EXERCISES

1. Once you have completed gathering the information you need, look back on the lists you created for the previous set of exercises and analyze what resources were most valuable. It may be helpful here to make notes about the successful strategies you used to gain good information. Why do you think they worked so well? What strategies did not work as well? Why not?

2. Find another writer in the company or in the class and exchange information on your research strategies. You may want to exchange the notes you've written for the exercise above, or you may want to simply discuss the strategies orally. After the discussion, write down at least one productive research technique you learned from this exchange of information.

CHAPTER 6

ORGANIZING THE INFORMATION

A documentation manager at a local software company is fond of saying, "People don't buy drills because they want the drills themselves; they buy them because they want holes. The same is true for computers." People buy computers to get things done, not simply to own high-tech machinery. And they certainly do not buy computers—or software—to read the accompanying documentation, although good documentation gives products a marketing edge. When computer manuals are well written and well organized, they work so smoothly that users hardly notice them: They don't get in the way.

Designing documentation that allows people to use the product without getting lost in the instructions is hard work. The process requires writers not only to understand the product thoroughly, but also to anticipate the users' needs and design a document that meets the varied expectations those users may have. Remember that the ultimate goal of computer documentation is to give users easy access to the technology without calling attention to itself. This chapter provides guidelines for organizing documents that are genuinely easy to use, not roadblocks to understanding.

═══GUIDELINES FOR ORGANIZING

Now that you have done all the preliminary research and have gathered sufficient information, you are ready to organize the material into a usable document. Naturally, the organizational plan you choose depends on the type of manual you are writing and on the needs of the user. Tutorials will be organized differently from reference manuals, and user guides differently from reference material. The bottom line is to think about what the user needs to do with the product, and organize your manual accordingly.

In general, however, it is a good idea to follow a few simple rules to make the document convey the information clearly and uniformly:

- Organize in task-oriented, sequential steps.
- Construct deductive frameworks.
- Organize similar types of information similarly.
- Separate instructions from expository prose.
- Include "user cues."

ORGANIZE IN TASK-ORIENTED, SEQUENTIAL STEPS

As you plan the document, determine the tasks the users must complete and the order in which they will complete them. If you are writing a hardware document, your job is easier because hardware tasks are most often discrete chronological steps. Organize the material to match the user's step-by-step tasks through the necessary procedures. If you are writing a software manual, on the other hand, you will discover that users may choose many paths to reach the same goal, and they may make different decisions about how to proceed. As a result, you must attempt to divide the various software procedures into separate tasks and organize the document around the most common task sequences users follow. For example, when writing a software reference manual, you probably will organize it by listing commands or by listing common tasks most users need to know. That way, people using your manual can mix and match the material as they need it. When writing a user guide, you must move sequentially through all the tasks the users must complete, from the most simple to the more complex.

One of the most frustrating things for users is to read manuals that have information spread out in so many places that they cannot complete one task without flipping to several different places in the documentation. This "GO TO" documentation (so named for the frequency of sentences similar to this one: "To know more about this procedure, go to Chapter 5…") is a major headache for anyone trying to complete a task. While it may not be possible to keep all the necessary information for every task in one place, the more often you can do so, the better your manual will be.

CONSTRUCT DEDUCTIVE FRAMEWORKS

Most people feel more confident in situations in which they know what to expect. Setting expectations for users is an essential element in good documentation. When users understand the nature of the task and its basic parameters before they begin, they can accomplish it much more quickly. As you write, make it a habit to explain where you are going and to give any other necessary context before you continue with specific details. For example, it is always a good idea to present the goal of any procedure before you actually list the steps:

> To print a document, follow these steps:
> - Go to the CM line by pressing [F5]
> - Turn on your printer
> - Type "ty" and press [F9]

This movement from general to specific information gives users a sense of context for what they are doing and allows them to understand the instructions more clearly. To reinforce this "deductive" organizational pattern, make sure you use it at every level of the document, from the large-scale overall organization to the design of the specific chapters and procedures.

ORGANIZE SIMILAR TYPES OF INFORMATION SIMILARLY

This rule is common sense, but it is often ignored in the heat of composing. In order to set user expectations and maintain uniformity throughout the manual, always present the same kind of information in the same way. For instance, all of the procedures should be formatted similarly and should use the same type of language, while all of the explanatory material should likewise be presented in a consistent manner. If you use numbers to indicate the separate steps in one place, do not switch to bullets in another. Even the specific terms used should be uniform: Do not say "Press [F1]" on page 15 and "Hit [F1]" on page 32. To create a professional-looking manual, be consistent in your design, format, and language.[*]

SEPARATE INSTRUCTIONS FROM EXPOSITORY PROSE

Most computer documentation contains both explanatory material and step-by-step instructions. It is important to keep the two separate so the

---

[*]Note that being consistent doesn't mean being dull. "A foolish consistency" may be the "hobgoblin of little minds," but creative consistency is both a challenge to the writer and a help to the reader.

users can easily identify the steps and quickly refer to them as they are per-forming the described task. If the procedures are buried in the expository prose of the paragraphs, users cannot follow them without stopping to read the entire paragraph.

One way to separate instructions from prose is to use numbered steps to describe procedures. If the order of the steps is important, use numbers to list the steps; if the order does not matter, you may use bullets or other devices. Again, be consistent in your usage.

It is also important to keep in mind several tips for writing instructions. When you separate the steps from the explanatory prose, you should also be careful to separate *user actions* from *system reactions*. The only thing that should qualify as a "step" is a user action. Anything the system does in response should not appear in the same format as a step, but instead should be clearly indicated as a response to the step. For example, the following procedure is written incorrectly:

> To call up a file:
>   1. Press [/].
>   2. The MAIN MENU appears on the screen.
>   3. Press [f] to select a file option from the main menu.
>   4. Press [r] to select "retrieve" from the sub-menu.
>   5. A list of files appears at the top of the screen.
>   6. The cursor appears on the first file name.
>   7. Move the cursor over the file you wish to call up and press RETURN.

To write it correctly, separate the user actions from the system reactions:

> To call up a file:
>   1. Press [/].
>   The MAIN MENU appears on the screen.
>   2. Press [f] to select a file option from the main menu.
>   The SUB-MENU appears on the screen.
>   3. Press [r] to select "retrieve" from the sub-menu.
>   A list of files appears at the top of the screen,
>   and the cursor blinks on the first file name.
>   4. Use the [↕] and [↔] keys to move the cursor over the
>   file you wish to call up, then press RETURN.

By numbering only the actions the users must take, you maintain consistency in your instructions and you give a better sense of how complicated the task is. When reactions are numbered and mixed together with the actions, the number of steps increases and the task looks much more difficult than it really is.

Typography, spacing, and format can also play roles in differentiating action from result. If the actions *look* different and are in different positions on the page than are the system reactions, users can easily tell them apart. In the example above, the steps are visibly separate from the results, clearly setting user expectations.

There is one final suggestion to make here regarding instructions. Research has shown that people have difficulty focusing on more than seven steps (plus or minus two) at a time. Therefore, as you organize the document, try not to exceed seven (plus or minus two) steps *for each task*. That may mean that you need to break down some of the longer procedures into subtasks to stay within the limit, but your manual will be easier to use and less intimidating as a result.

INCLUDE "USER CUES"

To help users find their way from one procedure to another, plan on including visual and verbal signals that serve as navigational aids. These essential "user cues" are so important to the documentation, the next section is devoted entirely to a discussion of them.

=== EXERCISE ===

Revise the following version of a procedure from a word- processing manual. Make sure you follow the rules for organizing discussed above.

EXERCISE

The Merge Print feature of QWIK Word Processing enables you to merge information from two files, so that the same document may be printed multiple times, varying information with each printing. This feature is useful for sending form letters to a number of people using a mailing list.

A Merge File (mailing list, for example) is merged with a Preformatted Document (form letter, for example) during Merge Print.

The Merge File (i.e., mailing list) to be "merge printed" with the Preformatted Document (i.e., form letter) may be created with Leading Edge Word Processing, dBase II, Lotus 1–2–3, etc.

The Preformatted Document (i.e., form letter) to be "merged" with a Merge File (i.e., mailing list) is a document created with QWIK Word Processing.

PREPARING A MERGE FILE

The Merge File is divided into records. A record is the total data for each entry in the file; the name, address and telephone number of an individual, for example. Records are divided into fields. Fields consist of specific data such as the name, address, or telephone number, etc.

A record created with QWIK Word Processing may be up to 200 characters in length. However, the entire record must be displayed on the same line. Adjust margins if necessary. (See THE FILING SYSTEM, Create A Document and THE FORMAT LINE for more information.) Fields within a record are separated (delimited) by a backslash (\). Each record is ended with the RETURN key (▲) return symbol). A Merge File created with QWIK Word Processing may be edited using all of the QWIK Word Processing features.

The following is an example of a record divided into fields:

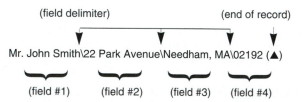

To prepare Merge Files using other programs, refer to the documentation provided with the program for the specific file formats required for that program. (NOTE: When preparing a Merge File for use with the QWIK Merge Print feature, the backslash (\) must be used to end a record. These delimiters must be included in your Merge Files.) Refer to the program documentation to sort or select records for merge printing.

Merge Files must be converted to an ASCII (American Standard Code for Information Interchange) file format for use with the QWIK Merge Print feature. (If the Merge File is created using QWIK Word Processing, See THE FILING SYSTEM, The Utilities Sub-Menu. If using another program to create the Merge File, refer to the documentation provided with that program for information on ASCII conversion. Always remember to specify the delimiters for fields and records.)

THE IMPORTANCE OF USER CUES

Effective documentation is not only well organized, it also reinforces that organization by providing highly visible signposts to guide the users. These "user cues" signal such things as overall organization, the hierarchy of user steps, the location of specific information, and so forth. With these cues, users can navigate more confidently through the instructions and can grasp the necessary information quickly. Without them, users are easily lost, mired in layers of information.

For example, notice how difficult it is to sort through the information in the following instructions:

RF-Fileserver 200/MC Installation

To avoid damaging the equipment, do not plug in the server power cord until the instructions in this manual indicate to do so. The dual fiber optic cable (ordered separately) connects the RF-H4000 transceiver and the RF-Fileserver 200. Each end of this cable is marked to indicate either the "transmit" or "receive" direction. To remove the protective caps from the server fiber optic connectors, unscrew the protective caps and remove them from the fiber optic transceiver cable by pulling them off from the ends of the fiber optic connectors. Never look into a fiber optic cable because the light emitted by the source may cause eye damage. Store the caps in a plastic bag taped to the cable. Gently press the half sleeves onto the ends of the connectors. To avoid damaging the fiber optic cable, do not touch the exposed fiber ends. To connect the fiber optic transceiver cable to the server, insert the receive and transmit fiber optic cable connectors into the input and output ports on the server. Observe the arrow direction on the cable. Assure that the knurled collar is not cross-threaded; tighten the knurled collar on the cable ends finger-tight. Do not overtighten the ends because the plastic fiber optic connector can be easily stripped. Verify that the other end of the cable is connected to an operational RF-H4000 transceiver.

Installers faced with this tangled prose will probably take twice as long as necessary to connect the fiber optic cables, and they are quite likely to damage both the equipment and themselves in the process. But with appropriate user cues added, the instructions become much more usable:

RF-Fileserver 200/MC INSTALLATION

CAUTION:

To avoid damaging the equipment, DO NOT plug in the server power cord until the instructions tell you to do so.

Connecting the Transceiver Cable

This section explains how to connect the RF-H4000 transceiver to the RF-Fileserver 200 using a dual fiber optic cable (ordered separately).

Before beginning, check to see that each end of the cable is marked to indicate either the "transmit" or "receive" direction.

Remove the protective caps from the server fiber optic connectors:

1. Unscrew the protective caps from the connectors.

WARNING!

Never look into a fiber optic connector because the light emitted may cause eye damage.

2. Save the caps in a plastic bag.

Remove the protective caps from the fiber optic transceiver cable:

1. Pull the protective caps off the ends of the connectors.

WARNING!

*Never look into a fiber optic cable
because the light emitted may cause eye damage.*

2. Store the caps in a plastic bag and tape it to the cable.

3. Gently press the half sleeves onto the ends of the connectors.

CAUTION:

*To avoid damaging the fiber optic cable,
do not touch the exposed fiber ends.*

Connect the fiber optic transceiver cable to the server:

1. Insert the receive and transmit fiber optic cable connectors into the input and output ports on the server. Observe the arrow direction on the cable.

2. Assure that the knurled collar is not cross-threaded.

3. Tighten the knurled collar on the cable ends finger-tight.

CAUTION:

*Do not overtighten the knurled collar
because the plastic fiber optic connector
can be easily stripped.*

4. Verify that the other end of the fiber optic cable is connected to an operational RF-H4000 transceiver.

THE CONNECTION IS COMPLETE

As you can see, the second version is easier to use because the installer readily can understand the hierarchy of information, the cautions and warning are highly visible, and the necessary actions are broken out into clearly defined steps. All of these techniques are user cues that provide a clear path through the information and give users familiar landmarks to increase their confidence in the instructions, thereby increasing the speed with which they can perform the tasks.

User cues come in many shapes and sizes, but are primarily divided into two major types: page format cues and textual cues.

PAGE FORMAT CUES

Page format cues include any design feature that makes it easier for the user to see how the information on the page is organized. For example: numbers, bullets, tabs, color, headings, running headers or footers, indentations, capitals, typeface, boxes, rules, columns, icons, white space, and so on. By using these techniques, writers establish hierarchies on the page—in other words, page formatting cues the user about how the information is grouped and how the various subgroups relate to each other. Further, page formatting aids users who must look away from the document to perform various tasks. These visual cues enable users to return easily to specific points on the page. They also help set user expectations about how to read the material because similar information will always be formatted similarly, thus making the document much easier to use. Users can scan the procedure before beginning to get an idea of its length and complexity.

User cues are also important in online documentation, although they may be quite different from those used for a static page. Chapter 14 discusses online cues more fully.

TEXTUAL CUES

Textual cues are actual prose elements within the text that guide the user through the information. For example: tables of contents, indices, overviews, introductions, conclusions, summaries, transitions, and so on. Any written material that gives the user navigational directions qualifies as a textual cue. In the fiber optic example above, the paragraph at the beginning that explains what the instructions are for and the sentence at the end that indicates the task's completion are textual cues. They signal users about what is coming, where to find information, and when the tasks are finished.

It's important for writers to realize how essential such cues are to documentation. Without them, users can easily become lost and frustrated. As you begin to plan the actual writing of your manual, think carefully about how the information should be organized and about how you can make that organization visible to the user. It may help to look at other documentation to see how other writers use these techniques—you may discover new ideas that you think will work well. Take some time to do some exploring before you make up your mind about what kind of cues to include in your document.

=== EXERCISE ===

Find a computer manual you think is easy to use and make a list of the user cues the writer has incorporated. Explain why you think these cues are effective.

═══CHUNKING

Imagine opening a novel and seeing nothing but continuous dense prose stretching from the beginning to the end of the story. Would you be interested in reading such a book? Probably not. Even in fiction, a solid block of prose intimidates most readers; in documentation, such uninterrupted prose is even more of an impediment. When users open a computer manual, they are looking for simple instructions, not long descriptive passages filled with monotonous information. At the other extreme, users are equally frustrated by manuals that contain never-ending sequences of steps with no indication of where one task ends and another begins. In both instances, the undifferentiated prose runs together into practically unreadable text.

Readers of any kind of prose—novels, poems, computer instructions, whatever—prefer to read text that contains information "chunked" into digestible units. When readers open such books, they can easily see by the arrangement of the text on the pages how to read the prose and where each part begins and ends. In documentation, this chunked information becomes especially important because users are constantly looking away from the manual to perform tasks, and they need to find their places again with minimum effort. Even more importantly, chunked prose allows users to identify at a glance the separate tasks they need to complete, thereby setting their expectations and giving them more confidence in their ability to do the job.

To make documentation more readable and legible—and therefore more usable—writers should plan ahead to divide each chapter into a series of digestible procedures or discrete chunks of information. If you are writing procedural steps, the "seven-plus-or-minus-two" rule serves as an effective reminder to divide long procedures into more manageable units. If you are writing more descriptive or theoretical passages, remember to group the information into a series of relatively small sections with clear subheadings. Even if you have an extensive amount of information to include in one chapter, it helps users if you break the material into a visible hierarchy of information by using different levels of headings and other user cues. This combination of more effective user cues and smaller units of information creates documents that are easy to read and easy to use.[*]

For example, read the following long passage filled with technical information. You will quickly see that the dense prose makes it difficult to grasp the ideas the writer is trying to convey:

---

[*]The controversy continues about whether to use a decimal numbering system for the document's subsections (1.0, 1.1, 1.2, and so on). While some hardware writers prefer to number the subsections, the trend is to number only the chapters, not the subsections. Some companies require numbering subsections—any documents done according to MIL-SPEC standards, for example, must be numbered.

You talk to your computer in BASIC, but your computer and your printer talk to each other in groups of digital pulses known as ASCII codes. In the ASCII code, each number from 0 to 255 has a particular meaning—36, for example, makes the printer print a dollar sign. Some numbers cause the printer to do other things, too. For instance, sending a 7 sounds the printer's bell.

Taken together, these numbers and their meanings make up the ASCII code (pronounced "ask-key"), which stands for the *American Standard Code for Information Interchange*. There are ASCII codes for all the letters of the alphabet (upper case and lower case), 0 to 9, most punctuation marks, and some (but not all) of the functions of the printer.

There are a number of different ways to represent an ASCII code, depending on how you are using it. For example, the ASCII codes for the letter "A" are 65 (decimal) or &H41 (hexadecimal).

BASIC uses the CHR$ function to represent ASCII characters and many functions. To print the letter "A" we would enter LPRINT CHR$(65). To make the printer's bell sound, we would enter LPRINTCHR$(7). In general, we print a character by entering LPRINT CHR$(ASCII code) to the printer.

We can also use hex ASCII codes. Although we use only decimal ASCII codes in this manual, you should understand at least what a hex code is. "Hex" is short for hexadecimal and refers to a base-16 number (the numbers we use in everyday life are base 10). Since the hex system needs 16 digits, it uses the numerals 0 through 9 and also the letters A through F. You can always tell that a number is in hexadecimal by the "&H" immediately preceding it. The ASCII code for the letter "A" (65 in decimal) is &H41 in hex.

Now look at the same passage "chunked" so that the user can clearly see the relative importance of each piece of information, can see how the all of the pieces fit together, and can more easily understand the material:

ASCII Codes and Your Printer

You talk to your computer in BASIC, but your computer and your printer talk to each other in groups of digital pulses known as ASCII codes.

What is ASCII?

In the ASCII code, each number from 0 to 255 has a particular meaning—36, for example, makes the printer print a dollar sign. Some numbers cause the printer to do other things, too. For instance, sending a 7 sounds the printer's bell.

Taken together, these numbers and their meanings make up the ASCII code (pronounced "ask-key"), which stands for the

American Standard Code for Information Interchange. There are
ASCII codes fro all the letters of the alphabet (upper and lower
case), 0 to 9, most punctuation marks, and some (but not all) of
the printer functions.

Decimal and Hexadecimal ASCII Representations

There are a number of different ways to represent ASCII code,
depending on how you are using it. For example, the ASCII
codes for the letter "A" are 65 (decimal) or &H41 (hexadecimal).

BASIC uses the **CHR$ function** to represent ASCII characters and
many functions. In general, we print a character by entering
LPRINT CHR$(ASCII code) to the printer:

Sample Decimal CHR$ Functions:

- To print the letter "A" in BASIC, we would enter
  LPRINT CHR$(65).

- To make the printer's bell sound, we would enter
  LPRINT CHR$(7).

We can also use hex ASCII codes. Although we use only decimal
ASCII codes in this manual, you should understand at least what
a hex code is. "Hex" is short for hexadecimal and refers to a base-
16 number (the numbers we use in everyday life are base 10).
Since the hex system needs 16 digits, it uses the numerals 0
through 9 and also the letters A through F. You can always tell
that a number is hexadecimal by the "&H" immediately preceding it. The ASCII code for the letter "A" (65 in decimal) is &H41 in
hex.

A word of caution is in order here. In your efforts to make documents
more readable, be careful not to go overboard and design text that is nothing but tiny chunks of information. Make sure that each division you make
in the text contains important enough information to warrant a separate
unit, and that the number of divisions is not so great that the text appears
scattered. Users must be able to see the hierarchy you establish; too much
division may destroy the clarity of the structure.

══════ EXERCISE ═══════════════════════════

Page through the computer manuals in your office, at home, in a computer
store, or wherever you can find documentation available. Look for examples of documentation that is overly dense or overly chunked. Choose one page and revise it so
that it communicates more effectively. Attach a photocopy of the page to your revision
so that it is easy to see the "before" and "after" versions.

▬▬INTRODUCTIONS, OVERVIEWS, AND SUMMARIES

Does this situation sound familiar? You are in the middle of assembling your new twelve-speed bicycle (or your new VCR cart, or the new picnic table for the deck...) when you suddenly realize you need a hexagonal wrench to finish the job. But you don't own one. Why didn't the instructions *say* you needed such a wrench! Now you have to stop everything and borrow a wrench from the neighbors or go to the hardware store and buy one. And to add to your frustration, you can't figure out what all those symbols mean in Step 2. You can feel the muscles in the back of your neck tighten as your head begins to pound. "Forget this!" you mutter as you slam down your tools. "The bike isn't worth this headache."

The pain of this experience could have been avoided if the instructions had clearly stated up front the tools necessary and had provided a key to the symbols used in the steps. Then you could have had the hexagonal wrench ready, could have breezed through the steps, and could have saved yourself a headache.

Fortunately, most computer manuals rectify this situation by beginning with sections called "Introduction," or "How to Use This Manual," or "Preface," or "About This Manual." Whatever its name, this section may contain the most important user cues in the whole document because it is here that the writer tells the reader who the appropriate audience is, how the manual is organized, what conventions are used throughout, and (especially in hardware documentation) what tools are needed. As the bicycle example suggests, these introductory elements are essential.

Right up front all documentation should indicate its audience and objectives so that users can tell immediately if the book is right for them. Users should not have to find out in Chapter 4 that they should be familiar with other manuals and other technologies in order to use this book. Nor should they get lost on page 25 because they do not understand the typographical conventions the writer is using. Does "Type: <ENTER>" mean the user should type just "ENTER" or "ENTER" plus the "< >" symbols? And, how can the user be sure which text is part of the instructions and which is an example of a system response?

By providing clear statements of audience and objectives and a "conventions" section at the beginning of the manual, writers go a long way toward preventing the user frustration (see Figure 6-1).

Other sections that are especially helpful for the user are the overview and summary sections. Some computer documents have entire chapters set aside to give an overview of the book, while others include overviews and summaries in each chapter. Some do both. Whether you choose to begin with an overview chapter and/or add overviews and summaries to each individual chapter, keep in mind that it is a good idea to give users a preliminary sense of the product and the manual they are about to use. As

Section Overview

This manual accompanies the Ciba Corning 780® Remote Data Manager® (RDM) and provides instruction for operating, maintaining, and troubleshooting the 780 RDM. This manual is part of a documentation set that includes the 780 *Remote Data Manager Quick Reference Guide* and *Installing and Customizing the 780 RDM*.

Who Should Use this Manual

This manual provides information appropriate for three types of users: laboratory supervisors, medical technologists or laboratory technicians, and laboratory pathologists.

If you are a...	Use this manual to...
laboratory supervisor	• perform data management tasks such as filing and archiving and system maintenance procedures such as backing up the hard disk and formatting diskettes for use
medical technologist or laboratory technician	• familiarize yourself with 780 RDM components and functions • perform daily activities using the 780 RDM
laboratory pathologists	• review the procedures for reviewing patient results and preparing patient reports

FIGURE 6-1   Example of Audience, Objectives, and Conventions Statements (Reproduced with permission of Paula Hammett, CIBA Corning Diagnostics Corp.)

Conventions Used in this Manual

Table 1–1 explains the conventions used in this manual.

Table 1–1. Manual Conventions

Convention	Meaning
ALL CAPITAL LETTERS	Any word or phrase that appears in all capital letters, bold type, corresponds to a screen message or menu option. For example, if the phrase 'Read 780 Data' appears as **READ 780 DATA**, it refers to a menu option on the Main Menu.
ALL CAPITAL ITALIC	Any word that appears in all capital letters, italic type, refers to a function key or other special use key on the keyboard. For example, if the word 'ENTER' appears as *ENTER*, it refers to the key board enter key.
• Bulleted List	A bulleted list contains a series of related items or topics.
1. Numbered List	A numbered list contains a series of action steps in a procedure.
a. Alphabetic List	An alphabetic list contains a series of sub-steps within a numbered list.
⚡ WARNING	This symbol is used in conjunction with warning statements. Warning statements provide electrical hazard information.
! CAUTION	This symbol is used in conjunction with caution statements, Caution statements provide information about personal injury hazards and product damage hazards.
■ NOTE	This symbol is used in conjunction with important statements. Important statements require your attention.

noted in the section on user cues, people appreciate knowing what is in store for them before they venture into unfamiliar territory. Having a deductive framework in which to understand the information presented puts users more at ease and makes their journey through the document more productive.

If you are writing a relatively short manual, including lengthy overviews is unnecessary. Likewise, in short chapters or documents, adding summaries is unnecessary because the user can remember all of the information clearly. Perhaps the chapter title, the headings, and the topic sentences will serve the purpose. But if the information you are presenting is complex and requires the user to integrate a number of procedures, overviews and summaries can clarify the process and prepare the user for the tasks ahead. The trick is to get the readers to conspire with you—these clues give readers a pattern to follow, and they will use that to locate the information. This technique increases their satisfaction while minimizing the work that they (and you) have to do.

Effective overviews can be as short as a bulleted list or a mini-table of contents on the chapter title page:

> This chapter explains how to
> - Define text.
> - Move defined text.
> - Copy defined text.
> - Save defined text.
> - Delete defined text.

Or, they can be introductory paragraphs that set the context for the material that follows:

> This chapter focuses on how to use the DEFINE command to high-
> light specified portions of your text. Once you have highlighted
> this text, you may choose to move it, copy it, save it, or delete it.

Similarly, summaries can be checklists at the end of chapters or of whole manuals, or they can be short paragraphs that remind the users what they should have learned by that point. In either case, they serve as a quick review of the information and as a reassurance for users that they have completed all of the necessary material.

EXERCISE

Write a brief overview and a set of conventions for the mail-merge document you revised for the first exercise in this chapter. After you have written these sections, decide if a summary is necessary. Will it aid the user, or is it "extra" information?

All of the organizational strategies discussed up to this point in the chapter are universal; that is, they apply to any documentation you may write. What follows are two particular organizational plans that work well under certain conditions. They are included here so that you are familiar with them and can use them if the material you are documenting lends itself to these techniques.

MODULAR DESIGN

Many technical writers have taken the idea of chunking information and have applied it rigidly to the overall organization of the document. In so doing, they have produced manuals that are "modular" in design. Modular design means that the entire document is divided into similarly formatted segments (modules) of approximately equal length; two to three pages is the normal length for each module. Each of these segments is usually a task the user must perform, and the entire instruction set for the task is displayed in a single module. In manuals designed this way, users can readily see the step-by-step procedures they need to follow, and they can also view each separate task in its entirety without having to turn the page. (Such design works especially well for online documentation because users do not have to move through multiple screens in order to finish a procedure. They move to the next instructional screen only after they have completed a whole task.)

In the example of modular documentation given in Figure 6-2, notice that the procedure spreads over two pages. This is the most common format. The running headers across the top of both pages give a clear sense of what the task is, and the complete set of instructions is visible at once. For most users, seeing the instructions in such a compact form is less intimidating than confronting a lengthy series of steps with no clearly defined beginning or end. This design inspires confidence—users come to expect information in the same order and arrangement and can tackle even the most complex material with greater ease than if it were presented in an unbroken linear fashion. As an additional benefit, modular design increases the manual's usefulness as a reference tool because the specific information appears in self-contained units that are easy to find and easy to read separately from the rest of the book.

Why, then, aren't all manuals designed this way? One of the main reasons is that not all computer information lends itself to division into equal segments, and a document that is half modular and half something else will serve only to confuse users. Some material is more easily conveyed in differing lengths and formats, especially information that is highly theoretical or that requires extensive explanation. In these instances, writers should still use chunking, but they need not be as rigid in its format as modular documentation requires.

Divide

Format:	divide *destination* by *source*
Variables:	The *destination* is a container and the *source* must yield a number.
Function:	Use the divide command to divide the value of *destination* by the value of *source*. The result is put into *destination*, replacing the old contents of *destination*. Any values in the *source* or the *destination* must be arithmetic expressions or numbers. See the **add** command in this chapter for a method of initializing variables before using the **add, subtract, multiply,** and **divide** commands.
Examples:	

divide field "SellPrice" by divider—divider is numeric value

divide line 2 of field "amount" by 3

▬▬ **FIGURE 6-2**   Modular Design in a Language Reference Manual

Another drawback of modular design is that it is expensive. When each module is separated from the others, the blank space in the manual increases and the number of overall pages increases. If the company is willing to pay for the extra pages, modular design may be an acceptable approach.

▬▬**INFORMATION MAPPING**

Sometimes known as "minimalist" documentation, information mapping is a design technique that reduces the number of sentences in the manual while increasing the number of tables and charts. The rationale behind this technique is that users do not have time to read long sentences. If they can process information in visual form more quickly, documentation should rely more on the visual than on the verbal.

In hardware documentation, it is relatively easy to depend primarily on graphic aids. In fact, some hardware instructions barely need words at all (see Figure 6-3).

But for software (and some hardware) documentation, words are often more essential. To convey the necessary verbal information so that users can process it visually, proponents of minimalist documentation use a series of tables like to the one in Figure 6-4. By using these tables, writers

DO

Format:	do *source*
Variables:	The *source* yields a text string.
Function:	Use the **do** command to get HyperCard to take whatever is in *source* and interpret it as a HyperTalk command. HyperCard can only execute one command at a time using the **do** command, so if there is more than one line in the *source*, only the first line is executed.
Examples:	

do field 3

do command—command is a variable containing a command string.

present a visual "map" of the procedure that allows users to grasp a great deal of information at one time.

Another instance when such a visual map is useful is to designate the choices users have at certain points in the document. Rather than writing long prose descriptions of each choice, using a "decision table" allows writers to present the options in a highly focused manner that is easy to read and uses minimum space (see Figure 6-5).

In prose, this information would take much more space on the page and users would not be able to understand it as quickly. As a result, many writers have tried to design their documents using a variety of these visual "mapping" techniques, from flow charts to elaborate tables with very few complete sentences users must read. Figure 6-6 illustrates a "mapped" procedure designed in a modular two-page spread. Tables are the primary element users see on these pages.

The benefits and liabilities of this technique are fairly clear. When information can be reduced to such lean formats, writers should do so, but that is not always possible. When the material requires more explanation, minimalist organization is not appropriate. However, unlike modular design, this technique can work well in tandem with other organizational strategies, as long as the same kind of information is consistently organized. In other words, your entire document does not have to be minimalist in design in order for you to use this technique where it is appropriate.

6.3.2 Floppy Disk Drive Cabling

Each floppy disk drive must be connected to three cables:

- Disk controller ribbon cable (includes two connectors, one per drive)

- Power cable (two provided, one for each disk drive)

- Ground wire (provided with disk drive kit)

Your AST 286 can accommodate up to two floppy disk drives (Figure 6-3)

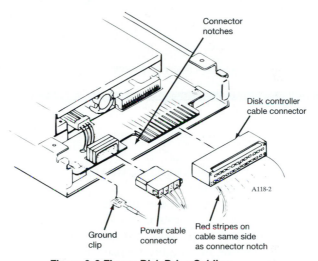

Figure 6-3 Floppy Disk Drive Cabling.

NOTE

Drive A must be connected to the connector at the end of the ribbon cable. If a second drive is added, it is connected to the connector located in the middle of the ribbon cable.

FIGURE 6-3 Hardware Instructions Relying on Graphics

FINAL THOUGHTS

Choosing the right organizational plan for your document is an essential part of the technical writing process. Whether you choose to use the strict modular or minimalist techniques, or whether you decide to depend on the

Figure 2–2. Circle Description Screen

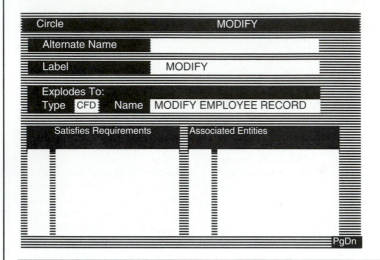

Circle	MODIFY
Alternate Name	
Label	MODIFY
Explodes To:	
Type CFD Name	MODIFY EMPLOYEE RECORD
Satisfies Requirements	Associated Entities
	PgDn

Table 2–1. Circle Attributes

Attribute	Length	Comments
Alternate Name	32	User's alias for this Circle. This name displays on selector lists.
Label	32	Text that displays inside this Circle on a Circle Flow Diagram.
Explodes To: Type	3	Three-letter code of the Circle Flow Diagram that provides a graphic view of this Circle: output only.
Explodes To: Name	32	Identifier of the Circle Flow Diagram that provides a graphic view of this Circle.
Satisfies Requirement: Type	3(×10)	Three- letter code of an Engineering or User Requirement associated with this Circle.
Satisfies Requirement: Name	32(×10)	Identifier of an Engineering or User Requirement associated with this Circle.
Associated Entities: Type	3(×10)	Three-letter code of an entity associated with this Circle.
Associated Entities: Name	32(×10)	Identifier of an entity associated with this Circle.
Description	72(×60)	Detailed, free-form definition of this Circle.

FIGURE 6-4 Information Table

4. Press *F2*.

A window menu is displayed.

Reviewing Plate Set-ups
(Continued)

If...	Then...
you want to print the current plate set-up	turn on the printer and verify that it is online. Select **PRINT PLATE SET-UP** from the window menu and press *ENTER*.
	A copy of the plate set-up is printed.
you want to delete the current plate set-up and return all the specimens on it to the worklist	select **DELETE, PARTIAL** from the window menu and press *ENTER*.
	This plate set-up is deleted and all information is restored to the worklist. The specimens from this plate set-up can then be included on other plate set-ups, held indefinitely, or deleted.
you want to delete the current plate set-up and delete all the specimens on it from the 780 RDM	select **DELETE, TOTAL** from the window menu and press *ENTER*.
	The plate set-up is deleted and all patient and specimen information from this set-up is removed completely from the 780 RDM.

5. Press *F2* when you finish reviewing plate set-ups.

A window menu offers two exit options:

- Select **SPECIAL FUNCTIONS** to return to

the Special Functions menu. Press *ENTER*.

The Special Functions menu is displayed.

- Select **MAIN MENU** to return to the Main Menu. Press

ENTER.

The Main Menu is displayed.

FIGURE 6-5 Decision Table

more general methods of information design, your primary goal should be to convey the information as efficiently as possible. Remember, users will follow only where you lead them. If you are vague or convoluted in your directions, they will get lost. Similarly, if you are inconsistent in the design

TORQUEING THE SYSTEM

Introduction	After the tubing and filter plates are installed, the filter cell must be secured using a torque wrench.
Important	It is important that the filter cell is evenly tightened. Therefore, the four hex nuts on the top of the filter cell are tightened gradually, using three different torque settings. Uneven torqueing can result in faulty seals (leaking) between the filter plates and the mixing of the filtered/non- filtered material. The torqueing procedure gives precise instructions.
Parts of the wrench	The torque wrench has the following parts.

Parts	Description
Head	The 3/8-inch-drive, deep-hex socket slips on and off the head.
Barrel	The wrench barrel is marked with measurements of torque. Torque is measured in units of inch-pounds and Newton-meters.
Sleeve Die	The sleeve die is marked with smaller increments of torque.
Lock Collar	The lock collar moves up and down. When the collar is pulled down, the barrel can be moved to adjust the wrench to the desired torque.
Socket	The 3/8-inch-drive, deep-hex sockets fits the 9/16 inch hex nuts on the cells.

Before you Begin	Before you tighten the hex nuts using the torque wrench, you should:

- install the washers and hex nuts on the four threaded extension rods of the cell
- simultaneously, finger-tighten the diagonal pair of hex nuts.

Continued on next page

FIGURE 6-6 An Information-Mapped Procedure in Modular DesignTorqueing The System

of your explanations, users will have a hard time following you. To keep them on the right track, give them a sense of where they're going at the outset and encourage their trust by providing frequent signposts along the way.

TORQUEING THE SYSTEM (Continued)

Torqueing Procedure

Follow these steps to torque the system.

Step	Action
1	Set the torque wrench to 40 inch-pounds (5.6 Newton- meters) by sliding and holding the collar back, and turning the wrench handle until the line corresponding to 40 inch- pounds on the barrel aligns with the zero reading on the sleeve die. Slide the collar forward.
2	Place the socket of the wrench over the first hex nut and rotate the the hex nut one-eighth of a turn to your right (clockwise). Do not turn the hex nut past this point.
3	Tighten the hex nut diagonally across from the first hex nut using the instructions in step 2.
4	Tighten the next hex nut (one of the remaining two that have not been tightened) as described in step 2.
5	Tighten the hex nut diagonally across from this hex nut using the instructions in step 2.
6	Repeat steps 2 through 5 until the torque wrench clicks at each nut position. *Note:* At the selected torque settings, the wrench will make a "clicking" sound. The wrench will not automatically stop torqueing if you continue to tighten the hex nut; therefore you must stop turning the nut when you hear the wrench click.
7	Set the torque wrench to 60 inch-pounds (7–8 Newton-meters) and repeat steps 2 through 6.
8	Set the torque wrench to 80 inch-pounds (10.2 Newton- meters) and repeat steps 2 through 6
9	If you are using the filter plates for the first time, wait approximately 30 minutes and then repeat step 8.
10	Repeat step 8 after every hour of system use to allow for compression of the separators.

Next Step

You are now ready to prepare your pump for operation. See next section.

FIGURE 6-6 An Information-Mapped Procedure in Modular Design (cont.)

═══ EXERCISE ═══

The following instruction set breaks nearly all of the rules for effective organizing. Referring as necessary to the List of Abbreviations, revise these instructions so that users can follow them without getting lost. Concentrate especially on designing the information so that the users' path through the material is clearly marked.

START-UP PROCEDURE

ON CPU

1. Set load unit address 354. (A00119)

2. Press LOAD button.

Operator will now be given the opportunity to insert or change the date and the time. If the time of the day is altered, remember to press the toggle switch TOD CLK.

(ABOVE ENTRIES WILL BE DONE ON THE OPERATOR MACHINE (061)).
Completion of IPL can now be done by entering 'IPL CMS' and press enter.

Sometime a normal IPL does not work. In these cases an IMPL should be attempted. Push the gray button on the CPU that is labeled Start Console File. This will load the IOCS onto Olf. Wait a couple of minutes until the message 'IOC LOADED AND OPERATIONL' appears on Olf. Then hit the load button on the CPU. The system should now come up normally. If it does not, then TECKO may have to be called.

When a display panel appears on the Operator console with autostart options for selected CMS machines, choose machines wish started by responding with the appropriate 'Y' for YES or 'N' for NO for the USERIDS you wish started and then press ENTER. The selected machines will automatically logged on.

From the operator console, enter the word 'READYN' and press enter. The READYN exec will cause all network lines and local control unit addresses to be enabled.

FINAL CHECK

A good practice is to check all network lines and local control addresses to verify that addresses are enabled.

It is a good idea to check whether RSCS and PASSTHRU are up. One can do this by entering the commands 'QRSCS' and 'QPASS' on Operator console.

Standard procedure for bringing up DOS can now be followed. The normal method for bringing up DOS is located in the DOS runbook which is located under device 060.

VS1B normally brings itself up. However, one should wait until the message 'VS1B is up and running' on Operator before actually logging on to VS1B. If one tries to logon to VS1B too early, it may get hung up.

List of Abbreviations Used in the Start-Up Procedure

CMS: Conversational Monitor System.

CPU: Central Processing Unit.

DOS: Disk Operating System.

IMPL: Initial Micro Program Load.

IPL: Initial Program Load.

IOCS: Input/Output Control System.

PASSTHRU: A TECKO program product; a networking facility.

READYN: The name of an executable file which resides in your CMS Library; it starts up (enables) machines in your network.

RSCS: Remote Spooling Communication System; a printing facility.

TOD CLK: Time-of-Day Clock; a toggle setting you use if you want to change the system date and time.

USERID: A number which represents one machine in your network.

VS1B: The name of your specific CMS system; VS is an operating system.

CHAPTER 7

AVOIDING LIABILITY:
Notes, Cautions, Warnings

No matter what product you are documenting, at some point you will need to emphasize the potential problems that can occur if the user makes an error. In software documentation, the problems are not usually life-threatening, although a slip of the user's fingers on the keyboard may crash the program or ruin a day's work. Hardware documentation specialists, on the other hand, often have the responsibility to warn users of physical danger. In both cases, poorly written warnings can have dire consequences for the user and for the company vulnerable to lawsuits. If there were a form of malpractice in technical writing, poor warnings would top the list of actionable items. In fact, companies have been sued over their documentation, and individual writers have been held liable—sometimes for what they did *not* say, as much as for what they said.

LIABILITY AND LANGUAGE

When writers publish information for the public, their words establish a kind of contract with the audience—a bond of faith based on truth and mutual understanding. Technical writers especially need to become more aware of this link and realize the multiple possibilities for their words to miscommunicate, thereby breaking faith with the audience. In corporate liability suits involving language, consumers usually claim intentional mis-communication or neglect on the company's part, while the company denies any such allegations. Somewhere in the transfer of information, the truth has become lost and the courts are left to decide who is to blame for the misunderstanding.

Language and the law is not a black-and-white issue. Writers cannot depend on a series of checklists to determine if their words meet the legal standards necessary to avoid liability. Instead, almost all litigation involving language is a matter of interpreting what was said, what was meant, and what was understood. Given the ambiguity of these situations, writers must sit down with their project teams and grapple with the questions covered earlier in this book: what is the nature of the audience, what is the nature of the message to be conveyed, and how can they adapt to each other?

Understanding how words act upon an audience is an important part of knowing how language can ultimately motivate the user to operate the product safely. It is essential that the cautionary information contained in computer documentation both communicates the problem and compels users to take the necessary precautions. If the warning or caution does no more than indicate potential harm but does not explain how to perform the procedure safely, it fails in half its purpose. Similarly, if a warning is present but not prominently displayed, it also fails to achieve its goal and the company could be liable for any injuries that may result. Technical writers should realize their responsibility to the users and to the company to write language that is clear, trustworthy, and unambiguous, especially when writing about potentially hazardous situations.

══WRITING EFFECTIVE WARNINGS

Cautionary notices to computer users generally fall into four categories: note, caution, warning, and danger, although "warning" is often used in casual speech to stand for all of these categories. As you write computer documentation, it is important to choose the type of notice appropriate for the information you are providing and label it appropriately:

- *Note* indicates information that might be of special importance to the user. For example:

    NOTE: The most common way to connect to a database is to use the autoconnect feature.

- *Caution* indicates information that the user needs to know to avoid damaging the software or hardware. For example:

> **CAUTION:**   An incorrect AC voltage setting can damage the server.

- *Warning* indicates information that is essential to people's safety. For example:

> **WARNING:**   Never look into a fiber optic cable because the light emitted by the source may cause eye damage.

- *Danger* indicates information that the user needs to know to avoid death or serious injury. For example:

> **DANGER:**   Unplug the computer before performing this procedure. You may be electrocuted if the computer is plugged in.

Once you have determined which type of notice to use, you then should think about how to design the notice so people actually pay attention to it and so that it complies with the legal duty to warn consumers. As you think about this design, remember that each type of notice should appear in a consistent format throughout the document, and users should see the notices *before* they perform the potentially dangerous procedures. Too often, warnings appear after the fact—an oversight that is the source of much injury and resultant litigation. Figure 7-1 shows a warning notice properly placed at the *beginning* of a procedure.

Although every company designs notices differently, certain key principles are necessary to create the appropriate effect on the user. Writers of effective warnings:

- Realize the legal basis of the duty to warn, including considerations such as the nature of the product, its use, the experience of the user, the frequency and seriousness of the potential injuries, who should be warned, the obviousness of the danger, the foreseeablity of misuse, and the feasibility of warning.
- Pay attention to any existing standards, either in-house or out-of-house, for writing warnings. (Some companies have guidelines published in their corporate style guides. The American National Standards Institute (ANSI) has established a set of standards to guide technical communicators in the development of effective hazard alert messages. (See ANSI Z535.)
- Consider the urgency, specificity, and clarity of wording.
- Place the warning in a location most appropriate for the user.
- Convey only one message in each warning. Multiple messages are confusing to the user.
- Give a clear reason for avoiding the hazard. The consequences of ignoring the warning should be spelled out in the notice.

2

2.6 Installation Procedure—Macintosh SE/30

                              WARNING:

We recommend that only qualified technicians install Mac SE/30 memory.
The Mac SE/30 system can produce a life-threatening electrical charge if the
installation is performed incorrectly, even when the system is unplugged. In
addition, the installation requires special tools. You assume the risk of damaging your machine. Apple warns that you will void the warranty if you
upgrade the SE/30 yourself.

1. **Assemble the special tools** (available from MacConnection 1-800-622-5472,
   MacWarehouse 1-800-255-6227, Programs Plus 1-800-832-3201, and other
   mail order suppliers):
   • Torx T-15 screwdriver with a long shaft of at least 5 inches (to remove the
     case screws)
   • Flat-blade screwdriver (to remove the programmer's switch and assist in
     closing the case back)
   • Case spreader (to separate the case back from the front bezel)
2. **Take appropriate anti-static precautions**. Use a wrist or heel strap to ground
   yourself.
3. **Shut the system down**. Select SHUTDOWN from the SPECIAL Menu. Turn
   off the power switch. Unplug the system, and disconnect the power cord,
   mouse, keyboard, and any other peripherals and cables from the computer.
4. **Remove the programmer's switch**. If the programmer's switch is installed on
   the side of the Mac, use a fingernail or the flat-blade screwdriver under the bottom edge to pry it off. (See Figure 9.)
5. **Remove the 4 screws**, using the Torx T-15 screwdriver. The Mac SE/30 has 2
   screws near the ports and 2 screws in the handle on top. (See Figure 9.)

FIGURE 7-1 Properly-Placed Warning (Used by permission of Clearpoint Research Corp.
35 Parkwood Drive, Hopkinton, MA.)

• Use as few words as possible so that the message is not surrounded by
  unnecessary clutter.
• Use the active voice, not the passive. The active voice expresses more
  urgency and is easier to read.
• Consider design elements that will make the message highly visible:
  icons, boxes, boldface type, color, pictures, and so on. Use these elements in a consistent manner. Figure 7-2 shows how graphics function
  in hazard messages.

⚠ WARNING: MOVING PARTS can cut off hand or fingers. DO NOT TOUCH.

⚠ CAUTION: HOT PARTS can cause burns. DO NOT TOUCH UNTIL COOL.

FIGURE 7-2 Hazard Alert Messages Using Graphics *(Source:* The Charles Machine Works, Inc., 1989. Used by permission of the Society for Technical Communication.)

As these principles indicate, careless use of language, misjudgment of the audience's needs, and even poor timing of information may result in ineffective warnings and possible litigation. To avoid miscommunication and liability, writers should take special care with the language they use to describe potentially hazardous situations. Where corporate attorneys are available, technical writers should submit for review all the warnings and cautionary notices they write. In the end, corporations bear the legal burden of proof that their writers' prose is trustworthy. Such ultimate responsibility requires responsible prose in the first place.

═══ EXERCISES ═══

1. Read the following hazard messages and decide if they are effective or ineffective. Explain your evaluations of each one. Then, try your hand at revising those you think are ineffective.

    (a) **NOTE**: Calls using the conferencing features of the Digital Conference Card (DCC) differ from the general call flow in several important areas. For a discussion of the conferencing call flow, refer to the Conferencing module of this series. Also, calls processed using the TeleRouter™ application overlay in an unhosted environment use a different call flow. Refer to the TeleRouter Reference Guide for more information.

(b) **WARNING**: To avoid damaging the equipment, DO NOT plug in the server power cord until the instructions in this manual indicate to do so.

(c) **DANGER**: As with most electrical appliances, electrical parts are electrically live even when the switch is off.

(d) **NOTE**: An object must be selected before any editing can be done.

(e) **CAUTION**: Each control lot number must be unique and should not be used for any other control for the selected test.

(f) **CAUTION**: Be careful when handling the toner cartridge. Spilled toner can be difficult to remove from clothes. If you do spill the toner, it should be removed with cold water; hot water may cause it to adhere to the clothing or skin.

(g) **WARNING**: The print head gets hot during operation, so let it cool off before you touch it.

(h) **NOTE**: Avoid using the RESET button while a data file is in use. If you reset your computer, the data entered prior to the reset may be lost. Data must be saved prior to reset. Check your application software documentation for instructions on how to save data. Reset also erases any data stored in RAM. If you are using a RAM disk, store data to either a floppy or a fixed disk prior to reset.

(i) **NOTE**: Express Publisher looks different when run on different types of graphics adapters. Don't be concerned if your document doesn't look exactly like the screen shots in the next two lessons. See Appendix B, "Performance Considerations," for more information.

(j) **WARNING**: Always wait until the amber light on the disk drive turns off before attempting to remove a diskette.

2.   Think about how you would design notes, cautions, warnings, and dangers for a computer manual you are writing. (If you are not writing a manual at the present time, find a sample manual and redesign the hazard messages it contains.) Produce one design for each type of message. Make sure they are sufficiently different so that the level of urgency is clear, but also make sure they look similar enough to fit consistently in the same manual.

CHAPTER 8

UNDERSTANDING TECHNICAL WRITING STYLE

Consider the following excerpt from a functional specification:

> The 752 controller may satisfy a read request via data in the read ahead buffer. If a read IOPB is satisfied by data already in the FIFO rather than a traditional disk access, the IOPB will be updated with the RAIOPB bit set (bit 0) in the 752 internal status byte (byte 3). This bit has been added as a performance analysis tool to accurately measure the hit-rate of the read- ahead offered by the 752.

Do you understand what it says? If you read it several times the meaning gradually begins to come clear, although the acronyms make little sense unless you are already familiar with the technology described. As a technical writer, you must find the engineer who wrote this specification, determine its precise meaning, and then translate it into prose readily accessible to the user. In doing so, you are presenting the same information but dressing it in a style that better fits the occasion.

Many writers are unaware that their style has an effect on readers. If the information is presented completely—no matter how hard it is to read—these writers feel they have done their jobs properly. For technical writers, such an attitude can undermine the quality and ultimate usefulness of the documentation they write. More than anything else, technical writers must be good *explainers*; that is, they must be able to take complex information and present it in a form that users can easily understand. Densely complicated prose used to describe complex information does more to block communication than to facilitate it. On the other hand, if writers pay attention to the stylistic elements that increase clarity, their documents increase in readability and usefulness.

STYLE *VS.* GRAMMAR

A common misconception some people have is that style and grammar are the same thing. They are not. The key difference between the two is that style is a matter of choice, while grammar depends on rules. For example, you can write "Four score and seven years ago..." or you can write "Eighty-seven years ago..." creating different stylistic impressions with the same information. Both are correct. But if you write "Our forefathers brung forth a new nation...," you have moved out of stylistic choice and into grammatical error.

Grammar and mechanics (that is, punctuation) follow rules that determine whether your prose is correct or incorrect. In school, you learned these rules from English handbooks and countless grammar exercises, but it was more difficult to learn how to make effective choices about style. When you write computer documentation, the rules of good English tell you how to write correct grammar, but you still have choices about diction and sentence structure that make a profound effect on your readers.

Diction is the writer's choice of words, while *sentence structure*, obviously, is the way the writer assembles the parts of the sentence. Under the heading of "diction" fall issues of synonyms, acronyms, formal vs. informal language, general vs. specific words, figures of speech, wordiness, and so on. "Sentence structure" includes decisions about sentence types, sentence lengths, syntax, and rhythm. There are very few, if any, handbooks or rules that tell you how to make these choices because each situation is different and therefore requires different decisions. As the audience and purpose of the writing change, so do the stylistic choices.

Many companies have established style guides to help their writers. In these guides, companies suggest uniform approaches to style that produce conformity in the corporate documentation. For instance, the style guide may determine how writers should handle acronyms and abbreviations, although these matters have no grammatical rules to govern their use. Suggestions about handling page design and other company-specific issues such as designing notes, cautions, and warnings may appear in the style guide. (Chapter 9 discusses style guides in detail and explains what sort of material appropriately belongs in them.) But even these guides do not cover all of the stylistic choices writers must make as they compose their documentation. The ultimate responsibility rests with the writers themselves.

=== EXERCISE ===

From the following list, determine which are style issues and which are grammar/mechanics issues. If some of them could be both, explain instances where they would fit in either category.

Abbreviations and acronyms
Capitalization
Commas
Dangling modifiers
Footnotes
Hyphens
Numbers
Parallel structure
Passive voice
Possessives
Pronoun-antecedent agreement
Since and because
Symbols and icons
That and which
Verb tenses

STYLISTIC APPROPRIATENESS

Imagine opening the manual for a brand new software package and finding this paragraph at the beginning:

> Before we begin learning about PolyWindows Desk, we should agree on a common terminology so that you will know what I'm talking about. Since you can't argue with a manual, I'm afraid you're stuck with my terminology.

Most people react poorly to such informality and misplaced humor. Rather than creating the impression of user-friendliness, the author of this paragraph has put off most users and compromised his credibility. In trying to write "interestingly," the author has violated one of the cardinal rules of effective manual writing: Never allow personality to dominate the prose. It is impossible to read the above passage without being overly conscious of the writer's voice. After even a few sentences, that voice becomes more irritating than helpful and overshadows the important substantive material the manual contains. Users may find themselves thinking, "Get *on* with it! Just tell me how it works!"

As this paragraph illustrates, some styles are inappropriate for the intended audience. Much the same as the excerpt from the functional specification given earlier is too technical and jargonistic for the general computer user, this passage is too "folksy." Good writers try to find a style that creates a middle ground between the two: a style that conveys information clearly while not calling unnecessary attention to itself. For example, this paragraph from a medical software manual presents technical information in a readable, yet professional fashion:

> The 780 RDM software has comprehensive reporting functions enabling you to design the chart, lab, and profile report formats your laboratory requires. You can build custom libraries of interpretive remarks and comments to add to individual reports. In addition, a patient profiling function allows you to generate profile reports from patient results stored in the computer. Batch printing and review functions facilitate efficient processing of patient reports.

Note that this passage does not use the pronoun "I," nor does it make any attempts at humor. The author's personality is downplayed so that the information the users need predominates. But the prose is not unfriendly or hard to read. This author directly addresses the users as "you" and uses words that an intelligent lay person can understand. The net result is that most users trust this prose and have confidence it will explain the software procedures clearly.

Even manuals designed for novice end users can manage to sound professional and user friendly at the same time. Notice the combination of these two qualities in this passage:

> If you have used another desktop publishing program, you probably won't have any trouble learning Express Publisher. Express Publisher follows many conventions established by other programs. If you are new to desktop publishing, you should read the whole manual in order. If you're more advanced, read through the descriptions of the chapters found below and decide what you need to read.

As is especially evident in this example, the key to finding the appropriate style is to treat the users with respect. Never "talk down" to users or patronize them by assuming they are computer illiterates. Terms such as

"Let's now try to initialize the disk" foster the sense that the author is similar to a first-grade teacher speaking to a group of six-year- olds. ("Now class, let's open our books to page 10.") Instead, use words that you would use when speaking to an intelligent friend who needs the information you can provide. Thinking of users as friendly colleagues allows you to develop a style that communicates well without patronizing or confusing your audience.

EXERCISES

Evaluate the style of these excerpts from computer documentation. Decide if their style is appropriate and explain your opinion.

1. This is another convention I use in this manual. If I want you to hold down the <SHIFT> key while pressing the <END> key, for example, I will write it as <SHIFT> <END>.

2. VAX BASIC is the version of the BASIC programming language that is installed on the VAX computer. You can access VAX BASIC in two ways:
   • As an ordinary compiler.
   • As an interactive system, the BASIC environment.
   All BASIC source files used with the ordinary compiler should have **.bas** as the filename extension. For example, the filename for a BASIC program might be **prog1.bas** where "prog1" is the filename and ".bas" is the filename extension.

3. This first chapter helps you overcome the confused feeling you get when you begin to use a new program. It explains the basic methods of exchanging information with the software program: what the words and pictures on the screen mean, how to choose the commands, how to use the mouse, and how to move around in a document.

4. Because of the real-time nature of the telecommunications applications and the unacceptability of service disruption, a system must be as fault-tolerant as possible. SDS-1000 system controller redundancy reduces the risk of a single-point failure and increases the reliability of an application.

5. You now know that when you call up an existing document and mess it up beyond repair, you can clear the document from the text window (the document in memory, not the one on the disk) by using **ABORT**.

TIPS ON STYLE

Though some technical writers do not realize the effect style has on their documentation, hiring managers do. The people responsible for hiring writers spend many hours reading applicants' writing samples and making decisions about who gets the interviews and who does not. It often surprises job candidates when the employee screening team pays more attention to writing skills than to technical skills. All of that training in hypertext applications or UNIX-based operating systems means very little if the candidate cannot write efficient technical prose. Take a look at what a team of hiring personnel at Data General said about the writing samples and résumés of four applicants for entry-level writing positions:

> "Nice applications background. Writing sample is unimpressive, though. She doesn't seem to know when to use a comma."
> "Procedural descriptions in Sample 1 read more like source code than prose—very abrupt. Often passive. A chore to read."
> "The first chapter of the sample manual is choppy, wordy, and passive. The text is hard to read because of all the passive voice."
> "I'm not thrilled with his writing. Chapter 3 is heavily passive and awkward. The structure is not always parallel. He tends to use jargon."

These comments reflect the attention industry pays to effective writing, especially writing that is easy to read at the sentence level. While it is hard to present "rules" for such a readable yet professional style, a few tips may be helpful. Keep in mind that these are general guidelines appropriate to follow most—but not all—of the time.

AVOID THE FUTURE TENSE

The use of "will" in computer documentation is unnecessary and does nothing but contribute to the wordiness of the sentences. It's important to remember that users do not usually read an entire chapter or manual before they begin a procedure. Instead, they read a step as they perform it. Therefore, for the users the action is occurring in the *present tense*. Do not write: "Press <F5> and the cursor will move to the command line" or "To complete your printer installation, select Setup (S) from the Control Panel, then the Printer (P) item. A list of installed printers will appear." Instead, be more concise and write: "Press <F5> and the cursor moves to the command line" and "To complete your printer installation, select Setup (S) from the Control Panel, then the Printer (P) item. A list of installed printers appears."

USE SECOND-PERSON PRONOUNS

Using "I" or "we" calls too much attention to the author(s). The standard for computer documentation is the second-person pronoun "you." While some old-fashioned manuals even avoid "you" in favor of the passive voice or "the user," most contemporary documentation uses "you" in combination with the imperative voice. For example: "If you make an error, press the <Esc> key," or, "To exit Spellcheck, press <F1>."

VARY SENTENCE PATTERNS

As in anything, repetition produces boredom. In documentation, the sentences should vary in their lengths and their structures so that the prose "flows" smoothly and doesn't seem monotonous or choppy.[*] Too many people believe that technical writing demands a style filled with simple short sentences that all begin with subject followed by verb. Prose like that is nearly unreadable. Instead of writing all one type of sentence, vary the way your sentences begin: use introductory prepositional phrases occasionally or begin with an adverb. Make sure, too, to alternate the sentence lengths sufficiently so that the prose doesn't produce a repeating rhythm that will lull users to sleep. The key to effective sentence design is to write sentences that keep necessary information clearly in the forefront but that do not repeat the same pattern continuously. Notice in the following example how the sentence patterns repeat, creating a staccato, choppy rhythm:

> You cannot supply the VMS-required arguments on the command line. You have to create a parameter file. The parameter file must contain the -db, -dt, -ld start-up options. You can add more startup options. These are just the required options. See Chapter 3 for information on how to create a parameter file.

With varied sentence patterns, this paragraph becomes much less of a chore to read:

> Because you cannot supply the VMS-required arguments on the command line, you have to create a parameter file. This file must contain at least the -db, -dt, and -ld start-up options, although you can add more. For information on how to create a parameter file, see Chapter 3.

In the second version, the sentences vary both in length and in the way they begin. By combining some of the shorter sentences and beginning the new sentences with structures other than the subject-verb pattern, the writer has linked the ideas more smoothly and has created a rhythm that is easy to read. Examples like this explode the myth that good technical writing is composed only of short, simple sentences.

---

[*]The exception to this rule is documentation that is to be translated by machine (for example, "Caterpillar Basic English"). Machine translations require simplified English with strict Subject-Verb-Object sentences and an extremely limited vocabulary.

TREAT LIKE THINGS IN LIKE MANNER

Users are most comfortable when information is presented in a consistent way. Although this advice may seem to contradict the previous tips on sentence variety, it really does not. In descriptive paragraphs and in other longer prose chunks, you should pay attention to sentence variety. But when you are writing specific instructions, users expect to see the procedural steps treated consistently:

> To locate a patient report:
> 1. Select **ACCESS PATIENT INFORMATION** from the Main Menu. Press <ENTER>.
>   The Access Patient Information menu appears.
> 2. Select **FILE FUNCTIONS** from the Access Patient Information Menu. Press <ENTER>.
>   The File Functions menu appears.
> 3. Select **SEARCH FOR REPORT** from the File Functions menu. Press <ENTER>.
>   The Enter Search Criteria screen appears. This screen contains the fields: PATIENT, SPECIMEN, and TEST INFORMATION.

In this example, all the steps start with verbs, the responses are all indented and begin the same way. The users can follow them quickly. By presenting information in this consistent fashion, you signal the user that these are procedural steps and that the process itself is a simple set of linked actions.

AVOID NOMINALIZATIONS

A "nominalization" is a noun made out of a verb or an adjective. When people write, they often turn verbs into nouns to make the prose more formal. Someone may write, "We had a meeting," instead of "We met," although the latter expresses the thought more directly and concisely. To make your prose more efficient and easier to read, try to use verbs to make your point, rather than nominalizations. Verbs are more powerful and more direct than any other part of speech. The list below shows how statements grow in strength as the nouns are replaced by verbs, or, at the very least, by the *infinitive* form of the verb ("to" + verb: "to collect," "to install").

- What is the point of determination for shutting down the system? (nominalization)
  What determines when we shut down the system? (verb)
- This program helps you in the collection and presentation of data. (nominalization)
  This program helps you collect and present data. (verbs)

- Follow the procedure described below for installation of the product software. (nominalization)
  To install the product software, follow the procedure described below. (infinitive)
- After completing the RF-Fileserver 200 installation, perform the installation verification. (nominalizations)
  After installing the RF-Fileserver 200, verify the installation. (gerund and verb)

USE THE ACTIVE VOICE

Although the passive voice is grammatically correct, writers often misuse it. Passives are appropriate when you want to downplay the agent—the person performing the action—in the sentence. For example, the corporate CEO who must announce company layoffs would probably say: "Four hundred employees will be let go in the coming month," rather than "I plan to lay off 400 employees next month." The first sentence doesn't assign responsibility for the action because there is no agent in the sentence to perform the action. The second sentence places the agent right up front where he or she can't hide.

In computer documentation, it is important to use the active voice in concert with the second-person pronoun "you." For example, you should say "Call up the Data Dictionary," rather than "The Data Dictionary should be called up." The first version is much more direct and easier to read. In the following examples, note how much more efficient the active voice versions are:

- When the task of batch processing is completed, the next project should be begun. (passive)
  Complete batch processing, then begin the next project. (active)
- The entity types unique to XL/Interface are described in this chapter. (passive)
  This chapter describes the entity types unique to XL/Interface. (active)
- A READ statement is used to assign to the listed variables those values that are obtained from a DATA statement. (passive)
  A READ statement assigns to the listed variables the values obtained from a DATA statement. (active)
- Drive B can't be referenced directly once it has been joined to another drive's directory. (passive)
  You can't refer to drive B directly once you have joined it to another drive's directory. (active)

Although it is important to use the active voice in most situations, the passive voice is appropriate when the doer of the action is unknown or unimportant. For example: "The 780/RDM stores results for each control when data from the 780 Fluorometer is read into the 780 RDM." In this instance, the machine is doing the reading and an active voice version of the sentence would use unnecessary words and probably sound silly: "The 780/RDM stores results for each control when the read/write head in the 780/RDM reads the data from the 780 Fluorometer."

In your own writing, prefer the active voice. It is easier to read and makes your prose shorter, more powerful and more efficient.

AVOID LONG STRINGS OF MODIFIERS

Another habit of style that impedes clarity is writing long strings of nouns, adjectives, prepositional phrases, and other modifiers. When you are aiming for precision, as technical writers are, it is easy to fall into the trap of using too many modifiers in one sentence. For example, note the noun string weighing down the end of this sentence: "This paper is an investigation into the information processing behavior involved in computer human cognition simulation games." It would be a lot more readable if the author had revised the sentence to separate the string (as well as get rid of the nominalizations): "This paper investigates information processing in computer games that simulate human cognition." Here are other examples of various types of strings that need to be revised:

- Appendix A gives the access control list entry structure definitions. (noun string)
  Appendix A gives structure definitions for entries in an access control list.
- Four data-type-dependent constants partition the range of the function. (adjective string)
  Four constants partition the range of the function; these constants depend on data type.
- The first matter of difficulty is the problematic task of assigning some kind of value to the amount of and nature of information that users need to have before they can complete the procedure. (preposition string)
  The first problem is to assign value to the amount and nature of information that users need before they can complete the procedure.

AVOID SEXIST LANGUAGE

Pronouns that specify gender are often offensive when used in situations that include both sexes. Sentences such as "The user can keep his documents on file" is inappropriate because it creates the impression that all users are male. Likewise, "The user can keep her documents on file" inappropriately assumes the users are all female. There are three ways to fix this problem:

1. Use the plural:
   "Users can keep their documents on file."

2. Use second-person pronouns:
   "You can keep your documents on file."

3. Include both genders:
   "S/he can keep her/his documents on file."
   (This solution is cumbersome and difficult to read. If possible, avoid structuring your sentences so that you must include both he and she.)

Another point to keep in mind is the possibility for sexism in the use of names. If you use names in examples, try to make them androgynous, such as Lee, Terry, Chris, and so forth.

EXERCISES

Revise the style of the following passages so that they are more concise, more efficient, and more readable.

1. Each tutorial exercise is written in easy to follow steps. The first column lists the action to be taken. It also shows a typical screen display of these actions. The second column shows the keystrokes needed for the accomplishment of the action. The last column gives further instructions for action completion. Entries in this column appear in all capital letters. They refer to menu selections. Words that are shaded represent text that must be typed using the keyboard.

2. Use the **send** command to send a message to any object when you need to avoid the regular hypertext message- passing hierarchy. Messages are normally sent by just putting the *messageName* followed by its parameter list on any line of script. However, such a message will be sent up the hierarchy in the standard way (see Chapter 1). The **send** command is needed to cause an object to execute one of its handlers when that object would not be in the natural inheritance path of the message sender. An example of this is when a calculated result in a field needs to be recalculated whenever a value changes in a separate but supporting field. In that case, the supporting field will send a **closeField** message to the calculated field whenever the supporting field is closed and the new result will appear. Normally, **closeField** would go from the closed field to its card and on up, bypassing other fields.

TIPS ON GRAMMAR AND MECHANICS

Although this is not a grammar handbook, a few tips presented here may alert you to some of the common grammar problems that have become the bane of technical editors. Remember, unlike issues of style, grammar and punctuation issues are governed by definite rules writers cannot ignore and still write correct prose. Four areas seem to give technical writers the most trouble: dangling modifiers, parallel structure, pronouns, and punctuation within lists.

DANGLING MODIFIERS

For a modifying phrase, clause, or word to "dangle," the writer has to place it so far from what it modifies that the meaning becomes unclear. Often, this error happens because writers are overly familiar with the subject matter and assume the readers will naturally understand what is meant. Such assumptions can create some hilarious prose. For example:

> Described as the most revolutionary new development in microcomputer software, you can now create customized hypertext applications.

In this example, *you* are the revolutionary new development, not hypertext. Why? Because the introductory phrase always modifies the noun or pronoun that follows it, in this case "you." The writer should have written:

> Described as the most revolutionary new development in microcomputer software, hypertext allows you to create customized applications.

But the writer probably was in a hurry and assumed the user could make the proper connections. The result is incorrect prose with an absurd literal meaning.

When you use any phrase, clause, or word to describe something else, make sure to place the modifier right next to what it modifies. The following examples taken from poorly written manuals show how dangling modifiers create ambiguous and often totally unclear meaning.

- These graphic elements are added to your Freelance set using the ADD menu. (incorrect)
  Using the ADD menu, you can add these graphic elements to your Freelance set. (correct)
- To see a button or a field's information box, the button or field tool must first be selected from the Tools menu. (incorrect)
  To see a button or a field's information box, first select the button or field tool from the Tools menu. (correct—note that the "you" is understood as the subject of the imperative "select."

- When using the **length** or the **offset** functions, or when referring to portions of text in chunk expressions, it is often necessary to know exactly what characters are being counted. (incorrect)
  When using the **length** or the **offset** functions, or when referring to portions of text in chunk expressions, you must know exactly what characters are being counted. (correct)
- Entries in the columns which appear in all capital letters refer to menu selections. (incorrect)
  Entries that appear in all capital letters in the columns refer to menu selections. (correct)

PARALLEL STRUCTURE

One way to make your writing easier to read is to craft sentences whose parts are balanced and whose form reflects their content. If you compose each sentence so that two or more elements sharing coordinate relationships are structured the same way, your prose will be much easier to read. Caesar's words were *"Veni, vidi, vici"* (I came, I saw, I conquered), not "I came, I saw, I was a conqueror." In keeping the coordinate elements of the sentence parallel in structure, he achieved a balance that increased the power and clarity of his words. When you write, you may be faced with more information than fits concisely into such short sentences. The longer the sentence, the more important it is to keep the components parallel so that the reader does not get lost in the prose's shifting directions. For example, this sentence contains a lot of information presented in a confusing fashion:

> You can connect to the schema holder database at start up, and then a connection can be made to the RMS database within PROGRESS by using the CONNECT statement, or combining any of the connection methods works as long as you connect to the schema holder database before the RMS database is connected.

When placed in parallel structure, the sentence reads much more smoothly:

> You can connect to the schema holder database at start up, and then connect to the RMS database within PROGRESS by using the CONNECT statement, or you can combine any of these connection methods as long as you connect to the schema holder database before you connect to the RMS database.

The following examples further illustrate how to use parallel structure effectively:

- You can then connect to the RMS schema holder and applications databases at run time or by entering a parameter file name. (not parallel)
  You can then connect to the RMS schema holder and applications databases at run time, or you can connect at other times by entering a parameter file name. (parallel)
- An array is a set of fields, all with the same type, length, and equally spaced apart from each other. (not parallel)
  An array is a set of fields, all with the same type, length, and spacing between them. (parallel)

THIS, THAT, AND WHICH

This, that, and *which* are three pronouns that cause all manner of difficulty for technical writers. "This" is a demonstrative pronoun that must refer to a specific noun. Used incorrectly, it refers to a vague idea that readers may have trouble pinning down. For example:

> You should be aware of certain fields, such as social security numbers, that always have the same number of characters in them. This is because the cursor automatically jumps to the next field when you enter a field that has the same number of characters as the previous one.

The pronoun "this" has no specific noun to which it refers. As a result, the user is left to guess. Does it refer to the fields? The characters? Actually, "this" refers to the whole idea presented in the first sentence of the example. Revised properly, the passage reads:

> You should be aware that certain fields, such as social security numbers, always have the same number of characters in them. This similarity causes the cursor to automatically jump to the next field when you enter a field that has the same number of characters as the previous one.

As you write, check to make sure you do not use the pronoun "this" unattached to a specific noun. Determine what the pronoun really refers to and be clear.

That and *which* are two other pronouns that cause problems. When used to introduce a clause or phrase, "that" makes the words that follow it *restrictive*; that is, the clause or phrase is essential to the identification of the item the clause or phrase modifies. It cannot be left out of the sentence. "Which," on the other hand, introduces *nonrestrictive* material that is not essential to the identification of the item modified. Nonrestrictive clauses and phrases should be set off by commas. For example:

- Chapter 2 explains the considerations that affect schema design. (restrictive)
  This chapter explains nonprivileged functions, which are available to all system users. (nonrestrictive)

In the first sentence, the restrictive clause is the only clue to the specific considerations explained in Chapter 2. The chapter discusses only those considerations related to schema design.

In the second sentence, the noun *functions* has already been clarified by the adjective *nonprivileged*. The information that all users can use the functions is helpful, but is not needed to identify which functions the chapter discusses.

PUNCTUATING LISTS

Writing procedures manuals and instructions requires the use of lists: lists of components, lists of options, lists of all sorts. Unfortunately, when confronted with all of these lists, technical writers often ignore the rules of good English they learned in school. Especially problematic is the colon. A colon directs the user's attention to whatever follows it: a list, a definition, an instruction, or important additional information. A few simple rules govern the use of colons:

- Use a colon when you use *for example, the following, follows,* or *as follows* to lead into a formula, line of code, or vertical list. For example:

  This version contains the following changes:
  - Databases can be verified on line.
  - Database backups can be verified at the page, segment, and set levels.

- Use a colon at the end of a sentence introducing a list if that sentence is incomplete without the items in the list or if the items are incomplete sentences. For example:

  Your system contains three elements:
  - Video screen
  - Keyboard
  - Printer

- Do not use a colon at the end of a lead-in sentence to a formal example, figure, or table; use a period instead. For example:

  Use:
  Figure 3-4 is a diagram of the PARTS database.
  Do not use:
  Figure 3-4 is a diagram of the PARTS database:

- Never use a colon after any form of the verb *to be*. For example:

  Use:
  These file types are excluded:
     .BAS
     .COB
     .DAT
  Do not use:
  The excluded file types are:
     .BAS
     .COB
     .DAT

- Never use a colon after a preposition. For example:

  Use:
  Copy the file to the following:
     Drive A
     Drive B
     Drive C
  Do not use:
  Copy the file to:
     Drive A
     Drive B
     Drive C

In addition to knowing how to use the colon properly, writers should also know how to use other punctuation rules applying to vertical lists. The following is a quick reference guide to the other rules:

- Place a period after each list element if one of the elements contains one or more complete sentences and the introductory material is a complete clause. For example:

  Each of the examples does the following:
  - Declares the parameters and the global symbol names.
  - Checks the return status for the value LIB$_INPSTRTRU. If this value is returned, you know that more than 30 characters were entered at the terminal and that the extra characters were removed.

- Do not use periods at the ends of list elements if each list element is a phrase or word. For example:

  The system comprises the following parts:
  - Memory management option
  - Disk and controller
  - Double-density diskette and controller

- Do not use semicolons or commas as end punctuation for items in vertical lists when the introductory material is not a complete clause. For example:

  Use:
  If you encounter a problem with the hardware, you can:
  - Try to fix it yourself
  - Call Customer Service
  - Order a new unit

  Do not use:
  If you encounter a problem with the hardware, you can:
  (a) try to fix it yourself,
  (b) call Customer Service, or
  (c) order a new unit.

EXERCISES

Revise the following sentences so that they use correct grammar and punctuation. Each sentence may have more than one error.

1. The **international style** is used when you want to sort text that may have ligatures, such as e or u. This is useful if you are sorting cards by text which is in a foreign language.

2. The best way to alleviate this problem is to rename the index fields. By doing this, matching fields from CDD/Plus can take the place of existing index fields.

3. Use the Priority command only if your output device is a printer, screen, or records images.

4. To complete many items, the cursor must be moved to the item and then the correct choice made from the pop up menu that appears.

5. The data limits for several chart types depend on the amount of text and the number of graphic elements in a chart, which includes:
   - titles,
   - headings,
   - notes,
   - labels,
   - legends, and
   - grid and tick marks.

6. Once you select a part of a system from the main menu, you will always see a second menu, which gives you the option of adding, changing, or deleting information concerning part of the system.

7. The next three sections are:
   - Basic database maintenance
   - Functions of adding new rows
   - Changing existing rows and deleting existing rows

8.   Proportional fonts reserve varying amounts of space for different-sized charac-
     ters. This produces attractive output, but it is very hard to line up the columns.

9.   The system manager has a reference book which lists every font on the 9700.

10. In addition to the 45 built-in functions in HyperCard, you can write your own
     functions using the **function** command handler as discussed in Chapter 5.

CHAPTER 9

DEVELOPING A STYLE GUIDE

In large companies, technical publications are often written by many different groups from various departments. At IBM, Digital Equipment Corporation, or Hewlett-Packard, for instance, not only is computer documentation produced by different groups, but the groups may not be in the same building or even in the same state or country. Nonetheless, it is important for the corporate image and standards of excellence to be recognizable in all of the documentation.

In small companies, the documentation department may consist of two or three people contributing to the various writing projects. With even two people working on the same manual, different writing styles emerge. To avoid continual conflict during the collaboration process, the writers usually hash out their differences at the beginning of each project, but questions continue to arise.

The solution for both of these situations is for the writers to develop corporate style guides.

THE IMPORTANCE OF STYLE GUIDES

Style guides provide common standards for all documentation in the company. Writers who have questions about most stylistic decisions can look to the style guide for answers. Editors look to the style guide as a sort of bible that enables them to maintain consistency among the written products sent to them for review. In disputes between writers, between writers and editors, or between editors, the style guide serves as an objective peacemaker: "That settles it. According to the style guide, we should use a bulleted list here, not a numbered one."

From the writers' point of view, these guidelines allow the freedom to concentrate on other aspects of writing, such as technical accuracy and organization of material. Writers also know that the stylistic decisions the editors make about the drafts they see are not arbitrary; they are informed and consistent decisions. Editors, too, no longer have to worry about defending their corrections. Editors without the benefit of a corporate style guide can spend most of their work time arguing with headstrong writers who disagree with their editorial comments. From both the writers' and the editors' perspectives, a company that spends the time and money necessary to develop a corporate style guide is a company that cares about documentation—that's good news for everyone involved in technical publications.

Style guides are important, then, for several reasons:

- They ensure consistency among the documentation projects.
- They give the writers a resource other than editors to answer stylistic and grammar questions.
- They give editors a standardized set of rules that explain many of their editorial comments.
- They reflect corporate concern for quality documentation.
- They are accessible to anyone in the company who has questions about documentation.
- They are "living" documents that can be changed as necessary to accommodate changes in audience, technology, and style.

HOW TO DEVELOP A STYLE GUIDE

As with most things worth doing, developing a good style guide takes a lot of care and patience. It is not something that can be done by one person working alone. Because style guides affect everyone who writes in the company, all those concerned with writing should have some input in the development process. When the development team handles the process well, everyone feels consulted and therefore more willing to accept the final document. On the other hand, if the process is handled poorly, many people may feel left out of important decisions and therefore may resist using the style guide at all. Remember, one of the main reasons for having a style guide is to unify company style, not promote factionalism and ill will. A few suggestions about how to proceed may help here.

The *first step* in developing a style guide is to form an advisory committee that includes representatives from all areas who will be affected by the document. Such a committee may include technical writers and editors from several different project groups, marketing writers (if they will ultimately use the guide), representatives from engineering (to ensure technical accuracy), people from customer support or another group that is in close contact with users, and any others who may be helpful.

Second, this committee should decide on the general components of the guide and hash out what would be helpful to incorporate and what would best be left out.

Third, the advisory committee should agree on a team of people responsible for drafting the guide. Keep in mind that the team should represent as many concerns as possible and not be made up of members from only one perspective on corporate writing. It should include writers as well as editors. If necessary, name a project leader, though some smaller teams work best when no one person is "in charge."

Fourth, the advisory committee in concert with the project team should establish a review process that encourages timely responses and allows input from all concerned. This process should ensure that the guide is circulated to reviewers during several phases of the drafting process and when the document is complete. There definitely should be a procedure established for reviewers to "sign off" in writing on the document each time they see it.

Fifth, the advisory committee and the project team should determine deadlines for the project.

Sixth, to the extent possible, the advisory committee should enlist company-wide support for the final document. Having assurances of support at the beginning of the project encourages people to take the project seriously and to offer more input than they perhaps would otherwise. If they know they will use the guide later, and if they know that the corporate management is behind the project, they have more of a vested interest in working productively with the team.

Seventh, the project team should handle the style guide development process in much the same way they would handle writing a computer manual: audience analysis, doc plan, reviews, deadlines, and so on.

Eighth, the advisory committee is the place where all disputes and other problems are resolved. This assignment of responsibility allows the project team to concentrate on writing the document rather than arbitrating differences of opinion.

With an advisory committee and a project team working together on the guide, the process can be inclusive and open to all opinions without getting overly bogged down in bureaucratic red tape or corporate politics. And although it takes a lot of work and cooperation, developing a style guide in this fashion ensures a final document worthy of respect.

═══COMPONENTS

However you decide to go about developing a style guide, one of the first things you need to ask is "What should the guide include?" This question is not as easy to answer as it sounds. Even a quick sampling of corporate style guides from computer companies shows that the documents are as different as the companies themselves. Perhaps the best way to handle a discussion of this issue is to suggest what a style guide is *not*. By process of elimination, you can then understand more clearly what a style guide is.

A company style guide is not a general English handbook. It is not comprehensive enough about language rules to serve in place of these longer, more scholarly volumes. Writers should always keep a good dictionary at hand (perhaps one recommended in the style guide), and should turn to the standard English handbooks for more detail on questions of grammar and style. The *Chicago Manual of Style* and *Words Into Type* are two commonly used references.

Likewise, a style guide is not a composition theory book. Although it may contain brief suggestions about audience analysis and invention of ideas, its main purpose is not to teach people how to write. Instead, it assumes people can write already and simply need to know how to write in a fashion suitable for the company's purposes.

While style guides are not substitutes for handbooks, grammar books, dictionaries, or writing theory, most of them include the following components in various forms:

> **Table of Contents**: The standard TOC format.
> **Preface**: Sometimes called "Foreword;" gives the rationale for the style guide and describes the intended audience.
> **How to Use This Guide**: Sometimes called "Overview" explains the structure and reading path through the guide; lists other standard reference books the company recommends, such as *The Chicago Manual of Style, Words into Type, Webster's Ninth New Collegiate Dictionary*, and so on.
> **Alphabetized List of Style Suggestions**: The heart of the style guide; includes topics such as "capitalization," "cautions, notes, and warn-

ings," "chapter and section titles," "colons," "commands," "commas," and so on.

Standard Company Terms: Includes acronyms and abbreviations the company uses; often includes dictionary-like meanings for the terms; lists company spelling of all terms and indicates preferred usage.

Index: Appears at the back of the book and is cross- referenced.

Page Formats: Exemplifies standardized page layout and design.

Legal Notices: Includes necessary general legal information as well as standards governing the use of trademarks.

Other components that frequently appear in company style guides but are more optional are the following:

Overview of Company Style: A brief discussion of general style issues such as organization, readability, word choice, and so on.

Artwork Conventions: Gives corporate presentation standards for logos, figures, tables, graphs, and other art.

Company Products: Lists the products the company currently sells.

Technical Publications Support Services Directory: Includes telephone numbers for departments that specifically offer support services; for example, media services, graphics department, corporate education, and so on.

No matter what components you choose, keep in mind that you must list them in a form that is easy for user reference. The most common format is to list the entries alphabetically, but to do this you must also remember to cross-reference any entry that may occur under a different name or may be clarified by information in another entry. Remember, too, to *use examples for all entries.* Simply telling readers how to do something is not enough; you must also show them. It is standard practice to use company-specific examples rather than more general ones. By using material that is familiar to your readers, you illustrate more clearly how the rules apply to the company documents, and you increase the likelihood that the readers will understand and remember the material.

SAMPLES

Figures 9-1 through 9-6 appearing at the end of the chapter indicate the wide variety of formats and components used in corporate style guides. Consider these as examples only, not as formulas for style guide design; they may serve to give you ideas of what you can do in your own company, but every situation is different and requires a guide tailored to the specific corporate needs.

Preface

This Style Guide is intended as a standard reference to terminology, word use, and other stylistic considerations for all ABC Company's employees called upon to write.

Chapter One defines the purpose of this guide, and discusses the general principles behind the rules found elsewhere in the book.

Chapter Two, Elementary Usage, describes basic principles of grammar, punctuation, word use, and style. Specific examples illustrate each principle.

Chapter Three, Principles of Composition, covers some general practices that apply to nearly any type of business writing.

Chapter Four, Word Choice, lists words and phrases best avoided, along with alternatives to their use.

Appendix A, Standard Terms, is intended to ensure the consistent use of terms widely used throughout the computer industry.

Appendix B, Protected Terms, lists all protected words and names an ABC Company writer is likely to encounter, as well as the proper procedure for using them.

Appendix C, Glossary, defines a selection of computer-related words and phrases.

FIGURE 9-1 *Sample 1:* Style Guide Introductory Material

EXERCISES

1. Make a list of all the people in your company who should serve on the style guide committee. Explain in detail why each person should be included.

2. Find a local company that has developed a style guide and interview employees who worked on the project. What were their procedures? What worked? What problems did they have? What advice do they have for others beginning the process? How many company sites did they have to include? What tools did they use?

3. Preliminary to developing a full-fledged style guide, try this practice exercise:
   • If you are in a classroom situation, divide into groups of four to six people. If you are in a corporate setting, put together a group of four to six people who are concerned with the company's writing style.
   • As a group, develop a table of contents for a style guide pertaining to your company's written material or the written material you produce in class (papers, reports, manuals, and so on).

FOREWORD

This guide presents XYZ Company's house style for technical manuals and courses.

Why do we need a house style? To help us better carry out our mission: producing accurate information that helps customers use our software successfully. Users ranging from inexperienced end users to high-level technicians must be able to find the information they need—and once they've found it, understand it. This calls for a style that is simple and direct.

Our style is somewhat informal: conversational without being folksy, free of excess verbiage and technical jargon. Along with our new format, it enables writers to speak directly to users—to tell them just what they need and no more. This style departs in many instances from our former style. Managers, writers, and editors should read this guide to see what has changed.

Courses and manuals should follow these style guidelines to ensure consistency across our published materials. XYZ has always placed a high value on editorial consistency. It's ever more important now as we implement an integrated, cross-platform product strategy.

These style guidelines should fit the majority of cases. (Style sheets prepared by the editors will provide supplementary conventions for a specific document set or curriculum.) But keep in mind that these guidelines are meant to foster clarity—never impede it. Like any guidelines, they require a certain degree of interpretation. *So use common sense.* If in a particular situation following a guideline would hide information or result in a clumsy construction, make an exception. Then clear it with the writing manager or editor.

Zoe Stevens
Director, Technical Communications

Harry Breitner
Director, Education Services

FIGURE 9-2 *Sample 2:* Style Guide Introductory Material

The TEK *Corporate Style Guide* is a complete guide to style for technical user information. Adherence to the style guide helps to ensure that the technical information for TEK products and services is consistent in style, organization, and terminology.

Use this guide as your first reference for style questions. If the information you need is not in this guide, then check the following sources:

- *The Chicago Manual of Style*
- *Webster's Ninth New Collegiate Dictionary*
- *Words into Type*

This guide also contains information relevant to internationalization and translation.

Audience

This guide has two major audiences:

- The technical publications community of TEK Corporation
- Third-party partners of TEK or other TEK customers

Structure of This Guide

The style guide is divided into three chapters and an index.

- Chapter 1 summarizes the major points of good technical style.
- Chapter 2 contains a discussion of a wide variety of style topics, arranged alphabetically.
- Chapter 3 lists symbols, abbreviations and acronyms, and other terms commonly used in TEK technical information.
- The index is a reference to the terms and topics in Chapters 1 and 2. Because the terms in Chapter 3 are arranged alphabetically, they are not in the index.

Conventions

The following conventions are used in this publication:

Convention	Description
...	Horizontal ellipsis points indicate the omission of material from an example. The information is omitted because it is not important to the topic being discussed.
. . .	Vertical ellipsis points indicate the omission of information from an example or command format. The information is omitted because it is not important to the topic being discussed.
italic type	Italic type sets off references to terms used as or singled out as terms. Italic type also indicates the complete titles of manuals and cross-references to further information.
boldface type	Boldface type indicates topics discussed elsewhere in the text.
dashes	In examples, a hyphen indicates both a hyphen and an en dash, and two hyphens indicate an em dash. For example:
	The length does not change—even if the line spacing changes.
	The version number is 4.2–1.
	In text, – signifies an en dash, and — signifies an em dash.
nn nnn.nnn nn	A space character separates groups of 3 digits in numerals with 5 or more digits. For example, *10 000* equals *ten thousand*.
n.nn	A period in numerals signals the decimal point indicator. For example, *1.75* equals *one and three- fourths*.

■■■■ FIGURE 9-3 *Sample 3:* Style Guide Introductory Material

APPENDIX A: STANDARD TERMS

The following list has been compiled in the interest of standardizing spelling and punctuation.

Term	Comment
alphanumeric	no hyphen
ANSI (American National Standards Institute)	note spelling
applications software	s on applications
background	one word
back up	verb, noun
back-up	adjective
bi-directional	hyphenated
bit-mapped	adjective
C	no quotation marks
calendar	note spelling
cannot	one word
data base	two words
descenders	note spelling
DIN (Deutsche Industrie-Normen; German Industrial Standards)	note spelling
disk	note spelling
down load	verb
down-load	adjective
end user	noun
end-user	adjective
foreground	one word
gigabytes (GB)	note spelling

FIGURE 9-4 *Sample 1:* Style Guide Section

media and medium

---

media and medium

Use *media* for both singular and plural forms with the singular verb form. For example:

If your media consists of only one volume, mount that volume and proceed to step 2.

If your media consists of two or more volumes, mount those volumes and proceed to step 3.

The media is packaged in protective material.

---

menus

Use the following guidelines when discussing menus:

- Use the verb *choose* rather than *select* when picking an operation from a menu.
- Use initial letters for the name of a menu; the term *menu* is all lower-case.

---

modifiers

Use the following guidelines for modifiers:

- Place modifiers carefully; their position affects the meaning of a sentence. For example:

    The program only reads the SYSTAT file.

    The program reads only the SYSTAT file.

    The first sentence implies that the program reads the SYSTAT file but does not process it. The second sentence implies that the program reads the SYSTAT file and no other file.

- Make sure that a phrase or clause is not a dangling modifier.

    Use

    To indicate that a statement is to be continued, end the line by pressing the F13 instead of Return.

    To continue a statement, end the line....

    These sentences make it clear that the user, not the line, is the subject of the verb *indicate*.

    Do not use

    To indicate that a statement is to be continued, the line is terminated with F13 instead of Return.

    In this faulty sentence, the line seems to be the actor indicating that the statement is to be continued.

TEK Internal Use Only

FIGURE 9-5 *Sample 2:* Style Guide Section

modifiers

- Avoid unnecessary or indefinite modifiers. For example, you can usually omit the following modifiers without loss of meaning:

activity	actual	appropriate
associated	currently	existing
fairly	much	properly
quite	rather	several
simply	suitable	very

- Do not use long strings of modifiers. For example:

Use	Do not use
Entry point descriptions for system services	System service entry point descriptions
Structure definitions for entries in an access control list	Access control list entry structure definitions
The AFC11 analog-to-digital converter provides the following features:	The AFC11 is a flexible, high-performance, multichannel analog-to-digital converter…
Multichannel capability. With AFC11, you can…. High performance. The AFC11 increases the…. Flexibility. You can use the AFC11 in several….	

In the original of the last example, the modifiers *flexible* and *high-performance* are too abstract to enhance the user's knowledge. The revision not only eliminates the string of modifiers but also clarifies the information.

See also **that** and **which**.

money

Monetary values are country specific. Use the following guidelines when discussing monetary values:

- Avoid reference to monetary values of products or services in user documents.

- If you use monetary values in examples, include a comment in the source file indicating the purpose of the example. If the document is localized, the translator can design an appropriate example using local currency symbols or values. For example:

Gender

- Paragraph mark

Use numbers only if you have more than 6 footnotes for the table (this would be most unusual).

Source notes   Use a source note to identify the source of a table.

Example   This table has both footnote and source note:

Course	Rating
Arithmetic*	Easy
Algebra	Moderate
Calculus	Difficult
Spanish	Easy
Russian*	Moderate
Chinese*	Difficult

*Spring semester only

Source: *Quimby College Catalog*

Gender

When not to use   In general, avoid gender:

Instead of this	Say this
Each user must chart his own course	*Each user must chart a course*

Instead of this	Or this	Say this
PAY uses *the employee's* pay group assignment to determine how frequently *the employee* will be paid, the dates on which *the employee* will be paid, and whether a time sheet is required for *the employee*	PAY uses *the employee's* pay group assignment to determine how frequently *the employee* will be paid, the dates on which *he* will be paid, and whether a time sheet is required for *him*	PAY uses *the employee's* pay group assignment to determine how frequency, pay dates, and whether a time sheet is required

Style Topics
1–22

━━━ **FIGURE 9-6** *Sample 3:* Style Guide Section

If you *absolutely cannot* avoid gender without clumsiness or unclarity, use male pronouns (*he, his, him*).

Don't use:

▼ Male-female combinations

▼ Just female

When to use Do use gender freely in scenarios and examples. Balance between male and female:

> *Sue Quimby runs a software shop. She...*
>
> *Fred Russell is a C programmer. He...*

Heads

Levels There are 3 levels of head: head 1, head 2, and head 3.

Heads are hierarchical:

▼ Don't use a head 2 without a head 1

▼ Don't use a head 3 without a head 2

When to use Use heads for text identifiers that must appear on the chapter TOC.

In a manual	In a course
Every chapter begins with a head 1 (this means no text or label before the first head 1)	Every unit begins with 3 standard labels: *Objectives; Topics; References*
Often you may want to begin with a head 1 such as *What this chapter is about*)	The first head 1 follows

Style

In a manual	In a course
Lead cap:	Initial caps:
Going into space	*Going into Space*
How to get started	*How to Get Started*
Example 1: a quasar	*Example 1: A Quasar*
What is a black hole?	*What Is a Black Hole?*
Exploring Jupiter's moons	*Exploring Jupiter's Moons*

CHAPTER 10

THE ROLE
OF THE EDITOR

This chapter will not teach you how to be a technical editor—that is the topic of another book—but will explain the role of the editor in computer documentation.

For years it was difficult to find anything written about technical editing; editing was taught primarily in journalism classes, and the technical editor relied on courses in grammar, *The Chicago Manual of Style*, and on-the-job training. Recently, people in the field of technical communication have realized how important editors are to their profession and have begun to pay closer attention to the editing process. Many articles, books, and conference presentations on the subject of technical editing have appeared in the last few years. Rather than surveying all of them, this chapter explores the editing process, talks about the dynamics of the writer-editor relationship, and offers some suggestions for writers in departments that have no editors.

THE LEVELS OF EDIT

In the past, the editor's task was divided into three jobs. The editor might edit primarily for organization and content, particularly on a first draft of a new manual. In this edit, the editor could suggest reorganization, style revisions, format changes, new headings, and additional content. Also working with this draft, the editor might correct grammar, spelling, punctuation, and other mechanical details to ensure that the typesetter would have a document that was clear and accurate in these areas. Finally, after the document had been typeset and the galleys or camera-ready copy prepared, the editor proofread for typesetting and typographical errors, and sometimes did a "production edit," checking for art placement and accuracy.

In 1981, two editors at the Jet Propulsion Laboratory (JPL), Robert Van Buren and Mary Fran Buehler, published an article identifying nine levels of edit.[*] This article caused many others to rethink the editing process and to recognize the many discrete editing tasks that editors perform. The levels of edit identified by Van Buren and Buehler are as follows:

1. The *Coordination Edit* is the document planning stage and may be done by an editing manager. During this administrative "edit," the editor or editing manager works with writers, writing managers, and technical project directors to agree on writing and editing budgets, scheduling for drafts, reviews, edits, and publication procedures. The writer will probably not start writing until after this coordination work is done.

2. The *Policy Edit* is the time when an editor may work with the writers to remind them of the corporate policies that may require, for example, a revised preface, the addition of a readers' comment form, a new order number for the manual, and so forth. This edit is often done in early stages of the writing and may occur before the first draft is completed.

3. The *Substantive Edit* is a thorough edit of the completed first draft and may result in recommendations for revision or reorganization. The editor checks for content, organization, logic, completeness, and coherence.

4. The *Integrity Edit* will probably be done on both a draft and a final copy of the text. This edit is a check to make sure that all the parts of the document are there. That is, if the text refers to Figure 1, the edit verifies that there is a Figure 1.

5. The *Screening Edit* includes a grammar check but focuses primarily on professional standards—seeing that the graphics are professionally prepared, for example, or that other parts of the document meet high enough standards to qualify as a professional publication.

6. The *Mechanical Style Edit* may be done on both a draft and a final copy of the text. This edit checks individual words for correct hyphenation,

---

[*]Mary Fran Buehler, "Defining Terms in Technical Editing: The Levels of Edit As a Model," *Technical Communication,* Fourth Quarter, 1981, p. 10–14.

spacing, capitalization, and consistency in style. For example, this edit makes sure that the writer is consistent in writing "Reference 1" rather than occasionally using "Ref. 1" or "reference 1."

7. The *Language Edit* is a full edit for grammar, punctuation, word usage, and related matters. This edit is done to the final draft, and may be done to earlier drafts as well.

8. The *Copy Clarification Edit* is necessary if the manuscript is to be typeset rather than published electronically by the writers. It ensures that all instructions to the typesetter are clear and accurate. This edit may be done by someone in the publications production department.

9. The *Format Edit* is done on the final, camera-ready copy or galleys to see how the copy looks on the page. The editor checks to verify that headings are in the correct typeface and spaced correctly, that column width is correct, and that graphics are printed clearly. Once this edit is completed, the document is ready for final production.

Although it is extremely helpful to be aware of the many tasks editors actually perform, few editors or editing departments have time to edit a document nine times. Thus, most editors combine several of these tasks to make fewer editing "passes" through the document. In time-pressured situations, when the entire writing and editing cycle is very short, some companies may rely only on thorough Language and Mechanical Style Edits, and the resulting documents may appear less professional than if they had been edited fully.

It is also important to note that in many companies, the word "edit" and the word "review" have different meanings. For some, an "edit" is an internal process that deals only with the formal editing levels discussed here, while a "review" is a process handled by people external to the project or the department. For example, an editor assigned to the project performs the edits, while engineers, marketing personnel, or other project writers may do reviews. In most cases, reviewers check the accuracy of the document's contents, rather than reading for style, mechanics, or other kinds of editorial concerns. Before reaching the final draft, a document may undergo several edits as well as multiple reviews. These processes work together to make the final version as error-free as possible.

══DEVELOPMENTAL EDITING

Today, many technical communications departments are able to eliminate external production vendors and produce the finished documents on in-house electronic publishing systems. Because this is true, some of the editing levels previously necessary have fallen by the wayside (the Copy Clarification Edit, for example). It is also true that as writers have more and more control over the entire publication process, the editing process has changed to meet the challenges of new technology. Editing online throughout the writing process is common practice in many companies, as is the coordination of editors and writers earlier in the document cycle. The stereotypic view of the editor as a "grammar cop" who sees the manual only at the end of the process is far from the truth.

Instead, editors are becoming involved in writing projects almost as early as the writers themselves. In some companies, editors attend the early planning meetings so that they are as aware as the writers of any mitigating circumstances or any special directions the writers must follow. When an editor is assigned to a project from the beginning stages, the relationship between the writer and editor grows from the outset as the two work together on the project rather than being placed in the more traditional adversarial roles. This "developmental editing" technique encourages a team approach to the project, allowing both writer and editor to share the sense of pride in a document well done.

To establish a developmental editing cycle for a document, you need to follow only a few simple rules:

- Include editors in the early stages of document planning.
- Allow editing intervention frequently during the composing process, rather than saving the editing totally for the complete draft.
- Complete as many of the levels of edit as possible—some may be worth repeating several times during the drafting stages.
- Provide the editor with all the reviewers' comments as often as the document is reviewed.
- Establish frequent editor-writer meetings to review the writer's progress and the editorial suggestions. If necessary, include the reviewers in these meetings.
- As discussed in Chapter 9, use a style guide that has been developed by a team of writers, editors, and other concerned people in the company.

When writers and editors work together from the outset of a project, they can avoid delays and frustrations caused by the extensive revisions often necessary when writers develop the text with no editorial input until the very end of the process. By making the editing process more efficient and less painful, developmental editing saves both money and egos.

═══ EXERCISE ═══

To see how editing is done in the corporate world, make an appointment for an informational interview with an editor working in user documentation in your company or in a local corporation. Be sure to ask about the complete editing procedures used in the company and about the budget and personnel constraints under which the editors work. Write the results of the interview in a paper that explains how this "case study" clarifies the role of the editor in technical communication.

THE DYNAMICS OF THE WRITER-EDITOR RELATIONSHIP

In an ideal world, the relationship between writers and editors is excellent. Writers do their best writing and accept the editors' expertise in matters of grammar, word choice, formatting, and notations to the typesetter. A good editor is a second pair of eyes and of ears, because a reader should be able to "hear" a text, to check for the graceful language as well as grammatical correctness. Further, an editor is more likely than the writer to know how to mark a text for typesetting.

Unfortunately, the relationship is often strained. Why? Both writers and editors sometimes view their opinions as inviolate, a perspective that is bound to get in the way of productive cooperation. Further adding to the problem are editing cycles that allow the writer to complete full drafts before the editors see the document. This lack of developmental editing begs for trouble.

There are a few ways out of this difficult situation. The first is for someone in authority—a supervisor or manager, if possible—to sit down with both writers and editors and spell out what each one's task is, leaving the writer room for creativity in the communication process and the editor room for authority in preparing the most professional manual possible.

A second step is the resolution of the "who's in charge?" issue. Companies may be set up with the writers as superior to the editors, or the reverse, making one group inferior and hence resentful. If the manager can resolve this situation so that the groups are equal in authority (including pay and hiring levels), the areas of disagreement are more likely to be resolved.

Behind any resolution is a conceptual issue. If the writers understand that they are fallible and do not have perfect knowledge in all areas of writing, typesetting, and production, they will be more likely to rely on the editor for careful marking of the document. Similarly, if the editors understand that the writers need room for creativity and problem solving in designing and writing the document and work to clarify and enhance the writers' style, the conflict may be lessened. Conceptually, both writers and editors are aiming at the best possible document for the specific audience and purpose identified in the doc plan. This shared goal should override any minor differences.

The following sections describe some things editors and writers can do to make each other's work more productive. The first part gives suggestions for writers, and the second for editors.[**]

TIPS FOR WRITERS DEALING WITH EDITORS

To make the editing process more useful, writers should prepare the editors for seeing their documents. Make sure you write a cover memo or fill out a preformulated cover sheet that attaches to the front of the draft you send to the editor. If your copy is being edited online, convey this same information electronically. In your cover memo, include answers to these questions:

- What is the document's objective?
- Who is document written for?
- Where does this document fit in the doc set?
- What stage is the document in?
- What special circumstances are necessary to know?
- What type of edit are you looking for?
- What type of edit are you *not* looking for?
- Who else is editing/reviewing the document?
- When do you need the comments?

In addition, always make sure to talk to the editors personally when you give them the document and write specific questions in the margins or embed them in the document in another typeface. Arrange a meeting with the editor to discuss the finished edit.

TIPS FOR EDITORS DEALING WITH WRITERS

As an editor, you should be sure you understand the kind of edit you are asked to give to each document. If the writer has not indicated this to you, *ask.* Then follow these guidelines to make the editing process most helpful to the writer:

- Do not review the software. The writer may confuse comments about the software with comments about the documentation. Instead, if you have problems with the software, speak to the appropriate developer.

[**]Thanks to Beth Britt and Amy Travis, BBN Software Products, for their help with this section.

- Keep the audience in mind. Remember as you read that your point of view may not be the same as that of the eventual audience of the document. Try to think about how the intended audience will react to the information.
- Keep the type of document in mind. Some documents are read straight through, from beginning to end, while others are sampled a section at a time. When you review, you will probably read straight through the document, but try to keep in mind how the document will appear to someone not reading in that sequence. For example, some redundancy may be appropriate when users read the sections out of order.
- Note good things as well as things that need changing. An occasional "Well said!" where deserved lets the writer know that the editor appreciates the work and is not just being negative.
- Write legibly. Being clear about what you want ensures that your idea will be incorporated correctly.
- Give explicit examples. Use specific examples of what you want. If you see something that is incorrect or missing from the document, say how to correct it or add it. Don't just say that it's wrong.
- Be explicit about questions. If you have a question, be sure to say what it is, rather than just putting a question mark next to the paragraph concerned. If you are not explicit, the writer may not know what you are questioning.
- Watch your tone. Words come across differently when they're written from how they sound when you say them. For example, a comment that may seem funny to you when you say it may seem inappropriate as a comment on paper. Similarly, exclamation points next to editors' comments often come across as insulting. A good rule of thumb is to avoid jokes, sarcasm, and dramatic punctuation.
- Explain your editing markings. If you are using a special coding system (such as using different colored pens or different typefaces for different types of comments), be sure to explain that to the writer.
- Treat the writer as a serious professional, especially if you expect that kind of response yourself.

Once you have finished editing the document, *talk* to the writer about your suggestions and other comments. Then, if there are any questions or unclear concepts, both of you can work them out immediately rather than wasting valuable time tracking each other down or proceeding in an inappropriate direction.

Remember, the purpose of editing is to create an accurate document that will help the customers use the product. Careful and complete communication throughout the review process is critical for creating good documentation.

=== EXERCISE ===

Look at the sample page from an edited manuscript (Figure 10-1). What problems with the editing do you see? Make a list of each editing/reviewing error and explain why you think it is problematic. You should be able to use the material discussed in the preceding section to help you find the errors.

IF YOU DON'T HAVE AN EDITOR....

While most writing departments in large companies have both a writing staff and an editing staff, many smaller companies exist with only one or two writers and no editors. Writers in such companies may rely on themselves to write, review, edit, produce, and copy the documentation. In most cases, this scenario creates much too much work for the writer and results in poor documentation. If you find yourself working under these conditions, you may want to look for other ways to get some editing help.

Peer review is a common option for people in such situations—one writer acts as editor of another's writing. An advantage of peer review is that it lets every writer on the project know what the other writers are doing. This process can minimize redundancy between documents; provide an opportunity for writers to share information, examples, and tips; and, because each writer must rely on the others, help to keep the comments constructive. In the best cases, the writers are equally skilled at editing and do a good job, assuming they also know how to mark up a text for production and other necessary tasks. The difficulty with this option is that it places people in the awkward circumstance of being peers—fellow writers—and editors, a job that necessarily requires some distance from the written text and the writer. Resentments over peer review comments may haunt a publications department for weeks at a time. Another difficulty is that a writer may not be a good editor; he or she may not know how to indicate to another writer what is wrong with a text and how to fix it. However, a peer review is better than no review at all, and you should use it if you need to.

The second solution is to hire *contract editors*. They are often available through groups of editors who contract themselves out, through general freelance editors' groups, and often through colleges and universities where faculty and graduate students do editing for a contract or hourly

Where's the title?

I really think you should think about where goes in...

put this first

Don't say this

Companies gain a lot by publishing accurate, readable documentation. Customers can better use the features of the software; the customer support staff can focus on helping customers with real problems, instead of describing what should be described in the documentation; and education services departments can use the information in the documentation to help design courses.

To create good documentation, writers need help from the rest of the company. One way writers elicit help is by asking others in the company to *review* the documentation, that is, to give feedback about what they've written. This document discusses:

- the components of a review cycle   *don't use "cycle"*

- good review habits

"This document should be read by you if"

You should read this document if you are ever asked to review documentation. The guidelines described here will help you give more effective review comments.

Review Cycles   *Don't use*

don't use

A review cycle is the process of distributing a document to reviewers and incorporating their feedback. Over the course of a release, several review cycles occur, each planned to correspond to a certain level of "completeness" of a document. Therefore, some reviews occur when a document is in an early, first draft stage, to make sure the document is headed in the right direction, while others are done as part of giving a document its final checks and corrections.   *?*

Why is this here?

don't use

wrong

hah!!

This section outlines the steps in a typical technical review cycle. The steps listed here are what writers strive for in a review cycle not all steps occur in all review cycles.   *don't use*

don't use

don't use

I hate this title !!!

The Art of Good Reviewing   **BBN Software Products**   1

FIGURE 10-1 Sample Page from Edited Manuscript

rate. All of these editors may bring a great deal of expertise to the job of reviewing the writing of someone they do not know.

The third solution is to insist on good, thorough *tests* of your document. These tests can be on site (quality assurance and alpha tests) or off site (beta tests). Tests will not uncover spelling errors, grammatical errors, style problems, nor can they mark a text for production. In short, they are not really edits at all. Nonetheless, they can discover prose you thought was perfectly clear that remains confusing to the users. Chapter 12 gives more thorough information on document testing.

There are also some self-help solutions to a no-editor situation. The first comes in a variety of *software packages* which flag misspellings, passives, multisyllabic words, and other inappropriate style and grammar. Although these software products are tireless at finding such problems, they are only surface editors and cannot do the full editing job. Some of their shortcomings include the following:

- They cannot read for sense or meaning. Thus, using only one- or two-syllable words, you could revise a paragraph to satisfy the software, but the prose could be totally nonsense.
- They cannot tell when a sentence that violates a programmed style or grammar rule is more graceful than one that conforms to the rules.
- They cannot mark a text for production. You may have software that does some of the markings for you, but ultimately you must know how to mark the text to tell the software how to do it completely.

A final note to those writers who are working alone or are one of two writers in a company. If you are the only writer, it is probably easy for you to carry all your style guidelines in your head and to be consistent with yourself. Once your company starts to grow, however, these guidelines will need to be formalized for ease and for consistency. During the rare times when you are not overly busy, begin writing down your own style guidelines so that they have begun to take shape when you bring in a second, third, and possibly fourth writer.

PROOFREADING

Whether you are a writer or an editor, it is important to learn and use the standard proofreaders' marks. You will use these in correcting your drafts, editing other people's work, and in preparing your document for production. Figures 10-2 through 10-5 list the proofreaders' marks and show how they are used to correct a text.

Proofreaders' Marks

A set of proofreaders' marks is used to indicate corrections or changes to software publications. With material on line-printer listings, circle the copy you wish to change, and mark the change in the margin. To ensure that a minor change is not overlooked, mark an x or a small arrow (←) at the right margin opposite the change.

If the change or correction is not clear, write a brief explanatory note in the margin. If text is changed drastically, deletes it, provide new text, and indicate clearly where to insert the new text.

The example following the list shows the use of proofreaders' marks.

Instruction	Mark	Mark in Text	Corrected Text
Delete	ℒ	the good word	the word
Insert indicated material	∧	the word	the good word
Let it stand	stet	the good word	the good word
Make capital	cap	the word	the Word
Make lower case	lc	The GOOD Word	the good Word
Set in italics	ital	The word is word	The word is *word*
Set in roman type	rom	the word	the word
Set in boldface	bf	the entry word	the entry **word**
Set in lightface	lf	the entry word	the entry word
Transpose	tr	the word good	the good word
Close up space	◠	the wo rd	the word
Delete and close up	ℒ	the woord	the word
Insert space	#	theword	the word
period	⊙	This is the word	This is the word.
comma	ˆ	words words, words	words, words, words
hyphen	=	word for word test	word-for-word test
colon	⊙	The following words	The following words:
Insert semicolon	;	Scan the words skim the words.	Scan the words; skim the words.
question mark	?	Is this the word	Is this the word?
exclamation point	!	This is the word	This is the word!
apostrophe	ˇ	Johns words	John's words
quotation marks	ℒ/"	the word word	the word "word"
parentheses	(/)	The word word is in parentheses.	The word (word) is in parentheses.
brackets	[/]	He read from the Word the Bible.	He read from the Word [the Bible.]

──**FIGURE 10-2** Proofreaders' Marks Demonstrated

superscript	$\overset{2}{\vee}$	$2 \overset{2}{=} 4$	$2^2 = 4$
subscript	$\overset{\wedge}{2}$	$H\underset{2}{O}$	H_2O
asterisk	✳	word ✳	word*
Start paragraph	¶	"Where is it?" ¶"It's on the shelf."	"Where is it?" "It's on the shelf."
Run in	⌢	The entry word is printed in boldface. The pronunciation follows.	The entry word is printed in boldface. The pronunci-ation follows.
Move left	⌐	⌐ the word	the word
Move right	⌐	the word	the word
move up	⌐	th**a** word	the word
Move down	⌐	th^e word	the word
Align	‖	the word ‖the word the word	the word the word the word
Center	⌐⌐	⌐ word ⌐	word

TEXT WITH ERRORS

MESSAGES

ABFNAM, Ambiguous functoin name

> Explanation: A lexical function was truncated too too few characters to make the function name unique.

> User Action: Re enter the command; Specify at least four characters of the function name.

> ABKEYW, ambiguous keyword Explanation: A keyword or qualifier name was truncated to too few characters to make the keyword or qualifier name unique.

> User Action: The rejected portion of the command is displayed between backslashes reenter the command; specify at least four characters of the keyword or qualifier name.

ABORT, ABORT

> Explanation: An attempted operation was aborted.

> 1.   The system retunns this status for an I/O request that was canceled before it was completed.

> User Action: The operating system does not normally display this
> message; user programs should be coded to detect and respond to the
> status return.

ambkey, keyword is an ambigous keyword

> Explanation: The indicated keyword has been truncated too to few characters to make it unique within its context. This message is associated with a status code returned from the VAX-11 Common Run Time Procedure Library.

> User Action: Correct the source program. Specify enough characters in thekeyword to make it unique.

FIGURE 10-3 Sample Computer Manual Text with Errors

TEXT WITH PROOFREADING MARKS

⌐ MESSAGES ⌐

ABFNAM, Ambiguous function name

> Explanation: A lexical function was truncated too too
> few characters to make the function name unique.

> User Action: Re enter the command; specify at least
> four characters of the function name.

ABKEYW, ambiguous keyword Explanation: A keyword
or qualifier name was truncated to too few characters
to make the keyword or qualifier name unique.

> User Action: The rejected portion of the command is
> displayed between backslashes reenter the command;
> specify at least four characters of the keyword or
> qualifier name.

ABORT, ABORT

> Explanation: An attempted operation was aborted

> 1.   The system returns this status for an I/O request
>   that was canceled before it was completed.

> User Action: The operating system does not normally
> display this
> message; user programs should be coded to detect and
> respond to the
> status return.

ambkey, keyword is an ambigous keyword

> Explanation: The indicated keyword has been truncated
> too to few characters to make it unique within its
> context. This message is associated with a status code
> returned from the VAX-11 Common Run Time Procedure
> Library.

> User Action: Correct the source program. Specify
> enough characters in the keyword to make it unique.

CORRECTED TEXT

MESSAGES

ABFNAM, Ambiguous function name

Explanation: A lexical function name was truncated to too few characters to make the function name unique.

User Action: Reenter the command; specify at least four characters of the function name.

ABKEYW, ambiguous keyword

Explanation: A keyword or qualifier name was truncated to too few characters to make the keyword or qualifier name unique.

User Action: The rejected portion of the command is displayed between backslashes. Reenter the command; specify at least four characters of the keyword or qualifier name.

ABORT, abort

Explanation: An attempted operation was aborted:

1.   The system returns this status for an I/O request that was canceled before it was completed.

User Action: The operating system does not normally display this message; user programs should be coded to detect and respond to the status return.

AMBKEY, keyword is an ambiguous keyword

Explanation: The indicated keyword has been truncated to too few characters to make it unique within its context. This message is associated with a status code returned from the VAX-11 Common Run-Time Procedure Library.

User Action: Correct the source program. Specify enough characters in the keyword to make it unique.

FIGURE 10-5 Sample Text Corrected

═══ EXERCISE ═══

Edit the following sample text about the .CHAPTER command (for DEC VAX Runoff) by using proofreaders' marks and other editing techniques discussed in this chapter. When you are finished with this exercise, you should have a copy of the original, a marked-up copy, and a clean copy with the changes made.

The .CHAPTER command has produced the above chapter heading with the chapter numbre automatically calculated. As ou can see by comparing the infup file (see Apprndix a) with this output, RUNOFF has utilized the information supplied as a text parameter in the..Chapter command as our chapter title. The chapter heading has been darkened with the bold flag,*. After printing out the heading, Runoff has spaced down three (3) lines and than produced the formatted text you are reading. Prior to the chapter command, several command were given which hold throughout the document : .FLAGS   ALL;   .AUTOPARA-GRAPH,.PAGESIZE   and .LAYOUT.   .FLAGS   ALL enables Runoff flags not initially available by default including the bold flag used to enhance the above chapter heading;   .Autoparagraph generates a new paragraph every time a line begins with one or more blank spaces; .PAGE SIZE was used to change the size of the printed page from the default value of 58 lines and 60 columns to 55 and 60. Finally, information provided in a .LAYOUT command has caused the uunning pages numbers to be printed out at the bottom of every pag except the first page of each chapter.   A .TITLE command has also been inserted after the .chapter command. The txtt parameter given was 'Runoff Example' and this label witl be printed out as part of the running head information in the uppoer left hand corner of each page except the first page of a chapter.

TECHNICAL GRAPHICS AND PAGE DESIGN

Except for the true literati, readers are nothing of the sort: they are *searchers*. To do their searching, they start out as *viewers* or *lookers*. They flip pages, scan, hunt and peck, searching for the nuggets of information that they need and that might prove valuable to them.

Jan White, Design Consultant[*]

As the computer industry pays more and more attention to ease of use and effective interface design, graphics are playing an increasingly major role in documentation. With the advent of the Macintosh computer, Microsoft Windows, XWindows, and other new operating environments, graphical user interface (GUI) has become a primary concern for many technical writers. How do you design screen displays so that the user can easily access the information without having to wade through thickets of prose?

Graphical interfaces present new opportunities for software companies, especially. Rather than simply upgrading products by packing in more features, these companies are focusing their development efforts on

---

[*]Jan White, "Color: The Newest Tool for Technical Communicators," *Technical Communication,* vol. 33, no. 3, (August 1991).

making the features they already have more accessible to users. Now, rather than "features wars," software companies are engaging in "interface wars," competing fiercely to produce the most enticing, intuitive electronic interface in order to capture the popular computer market. With so much corporate money pouring into these designs, it's no wonder that most people associate technical graphics only with user interface. But the importance of good design goes well beyond online documentation. When writers produce hardcopy manuals, they must use graphics effectively *on the page,* without the help of electronic animation.

Chapter 14 discusses graphics for online documentation. The province of this chapter is the more basic elements of technical graphics for hardcopy: flow charts, screen shots, line drawings, and so forth, along with the principles of page design and the general visual presentation of documents.

▬WHEN TO USE GRAPHICS

One of the first questions writers may have as they begin to design their manuals is when to use graphics in their texts. Generally, three reasons for using graphics apply: Graphics increase (1) *efficiency* (by requiring less time to read), (2) *effectiveness* (by enhancing reader understanding), and (3) *impact* (by strengthening the reader's impression of each idea). For example, if you want to tell a novice user how to insert a disk into a disk drive, you know almost intuitively that your instructions will be clearer and that the user will learn to do the task faster if you accompany your instructions with a drawing. The same principle applies to many portions of your document. Whenever an illustration will enhance the efficiency, effectiveness, and impact of the prose, include it.

Once you have decided to include graphics in your manual, it is important to keep in mind that graphics should complement the written text, but the user also should be able to understand the graphics fairly well without the help of written explanation. Each graphic should have a clear caption indicating what it represents, and the drawing itself (or photo, chart, whatever) should be self-explanatory. Likewise, the prose in the text should certainly refer to the graphic, but it should not totally depend on the visual aid for clarity. When prose and graphics work side by side in this fashion without becoming interdependent, they reinforce the information and make the reader's job much easier.

▬KINDS OF GRAPHICS

Computer documentation can include the following kinds of graphics, as well as others: photographs, line drawings, bar graphs, line charts, pie charts, flow charts, screen displays, schematics, rules, boxes, bullets, and other symbols or icons. This section shows examples of each and briefly discusses when they are most appropriate.

Photographs may be either color or black and white and are usually pictures of hardware or display screens. You most often find photographs in brochures, annual reports, and other marketing pieces, as well as in hardware installation and troubleshooting manuals where the reader needs an exact representation. Although photos sometimes appear in instruction manuals, they are often not effective because they include too much background and incidental visual information rather than an uncluttered focus on the object or part in question. But in marketing pieces and occasionally in user manuals, photographs enhance reader interest and serve the important function of showing the product in particular settings. Often, marketing photographs also include people using the product—a real boost to the human interest angle. (Remember, though, that photographs of people may go out of date quickly because of the ever-changing styles of hair and clothes.)

Photographs also may convey attitudes as well as information—for example, a woman seated at a terminal and the man, presumably her boss, standing in the instructive, dominant position.

Note the two photographs in Figures 11-1 and 11-2. As you can easily see, Figure 11-1 would not suit a user manual because the product itself is

FIGURE 11-1   (Used by permission of GRiD Systems Corp.)

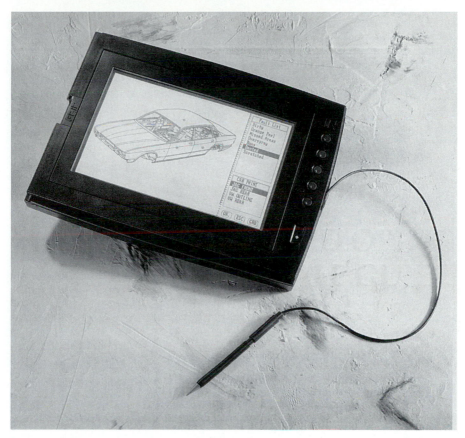

FIGURE 11-2   GRiDPAD$_{HD}$ and GRiDPAD$_{RC}$

The GRiDPAD$_{HD}$ and GRiDPAD$_{RC}$ are two models in GRiD System's line of pen-based computers. The GRiDPAD$_{HD}$ has a 20MB hard disk drive and one RAM card slot while the GRiDPAD$_{RC}$ has two RAM card slots for up to 2MB of storage. Both computers have a backlit, transflective display. (Used by permission of GRiD Systems Corp.)

not sharply in focus. Instead, the photo emphasizes the product's context: the well-organized office, the professional people, and the satisfied customer mentality. Figure 11-2 on the other hand, might work as part of a user manual because it clearly illustrates the parts of the machine. It is uncluttered by its context and the focus is only on the machine.

Line drawings are the preferred art in user manuals. They are particularly efficient because the artist can eliminate unnecessary detail and represent only those aspects of the product the user needs to see. They are also easy to produce and scan especially well into files for use in desktop-publishing systems. In addition, line drawings can be effective in depicting a sequence of actions, as the drawings in Figure 11-3 illustrate.

If you are connecting to thick Ethernet cable, push the switch to its opposite up position. This turns off the card's internal transceiver. When the internal transceiver is turned off, you can connect the Macintosh II to an external transceiver on thick Ethernet.

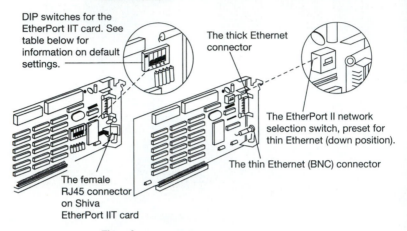

DIP switches for the EtherPort IIT card. See table below for information on default settings.

The thick Ethernet connector

The EtherPort II network selection switch, preset for thin Ethernet (down position).

The thin Ethernet (BNC) connector

The female RJ45 connector on Shiva EtherPort IIT card

Figure 2
Shiva EtherPort II network selection and IIT DIP switches.

The Shiva EtherPort IIT card has four DIP switches that come set for most network situations. Verify that the switches are in the default settings. The table below describes what each DIP switch does.

DIP Switch #	What it means	Default setting
1	Reserved	OFF
2	Signal Quality test Enable (SQE or heartbeat)	OFF (Enabled)
3	Link Test Enable	OFF (Enabled)
4	Reverse Polarity Correction Enable	OFF (Enabled)

Table 1
DIP Switch settings for the Shiva EtherPort IIT (10BaseT) card

FIGURE 11-3 Line Drawings (Used by permission of Shiva Corporation.)

Quarterly Sales Performance

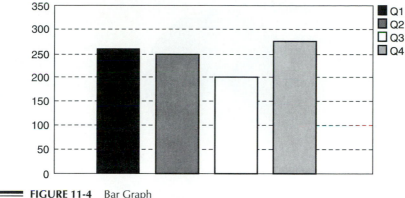

FIGURE 11-4   Bar Graph

Good line drawings can enhance a manual's efficiency, effectiveness, and impact. To prove this to yourself, try following a simple set of instructions without the graphics, perhaps by covering them up. You will probably discover how important the pictures are.

Bar graphs, line charts, and *pie charts* are simple graphics designed to show comparative or contrasting data. While these are rarely used in instruction manuals, they are a significant part of marketing writing and of software products for business presentations. It may help to know the function of each type of chart (See Figures 11-4 through 11-6):

FIGURE 11-5   Line Chart

Sales Volume vs. Profit Margin

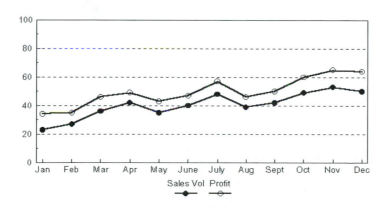

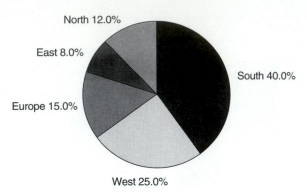

Sales by Region

North 12.0%

East 8.0%

Europe 15.0%

South 40.0%

West 25.0%

FIGURE 11-6 Pie Chart

FIGURE 11-7 Table

		Table 5-1 Status LEDs/Indications		
LED Name	**LED Definition**	**State**	**Indication**	**Corrective Action**
D1	Power ON/OFF	ON	The server's dc voltages are correct.	—
		OFF	The server's dc voltages are NOT correct.	Go to Sections 5.3 and 5.4.1.
D2	Diagnostic	ON	Self-test passed.	—
		OFF	Fatal error if LED remains OFF within 2 minutes after power-up.	Call Digital Customer Service.
		Blinking	Nonfatal error.	Go to Section 5.4.3.
D3	Software	ON	Server image successfully loaded.	—
		OFF	Down-line load in progress.	—
		Blinking	Multiple-load failure.	Go to Section 5.5.6.
D4	Network activity	ON*	Indicates activity on the network.	—
Fiber Optic Indicator LED	Fiber optic board power	ON	DC power to fiber optic converter.	—
		OFF	Problem in fiber optic converter circuit.	Call Customer Service.

*Can be ON, OFF, or flickering, depending on the amount of traffic on the network.

146

Bar graphs: Show changes in data over time at discrete time periods.

Line charts: Show changes in a data set over time, especially when the data set has many points.

Pie charts: Show parts of a whole.

Tables appear frequently in computer manuals. Simply defined, a table presents text and numbers without graphic interpretation. They let the facts speak for themselves. Tables are usually reference or summary tools and work well in manuals to present the information in a handy, easily accessible form (See Figure 11-7).

Another kind of table that is useful in computer documentation is the *decision table* (See Figure 11-8). This kind of table is an efficient way to indicate the choices users have. There are several types of decision tables, but all indicate user choices. One type follows the "If...Then" format. If they want to do X, they should follow Procedure Y; if they want to do A, they should follow Procedure B. Other types of decision tables graphically list all the information in a chart, allowing users to determine the appropriate coordinates of their specific situations. By collecting all of the options and placing them in an easy-to-read table, you avoid lengthy prose explanations and unnecessary repetitions.

Flow charts graphically depict a sequence of events or steps. Often, when you have a long series of steps the user must perform, a flow chart serves as a kind of quick-reference map through the procedure. By looking at the flow chart, the user can see at a glance how all of the steps relate and where each subtask is going. Flow charts are also helpful in illustrating how material is organized and how the parts of a whole work together (see Figure 11-9).

Screen displays in manuals serve to reassure users that they are operating the software correctly. For instance, if a user is following a step-by-step procedure, at significant points the screen display in the manual should match what is displayed on the actual computer screen. If it does, the user has performed the tasks correctly. If it does not, she knows to try again. Another use for screen displays is to annotate the information on the screen for the user. In other words, if you need to give brief explanations about items on the screen, a screen display printed on the page allows you to label it and quickly explain its parts (See Figure 11-10).

Schematics most often appear in hardware maintenance and repair manuals used by people in field or those working in manufacturing shops. These engineering drawings must be absolutely precise and should be prepared by someone trained in their design or trained to create computer-generated schematics (See Figure 11-11).

Rules, boxes, bullets, and sidebars enhance the readability of information, and make the pages look more attractive. Everyone realizes that a dense page of uninterrupted text is more readable if important information is set

1	If...	Then...
	you want to edit the highlighted report	use the arrow keys to highlight the desired patient and press *ENTER*. The Patient Test Result screen is displayed for the selected patient. Proceed to step 3.
	you want to return to the Access Patient Information menu	press *F2* to display a window menu. Select **ACCESS PATIENT MENU** and press *ENTER*. The Access Patient Information menu is displayed.
	you want to return to the Main Menu	press *F2* to display a window menu. Select **MAIN MENU** and press *ENTER*. The Main Menu is displayed.

2

**Table 2: Memory Module Installation Chart
(as you face the front of the system)**

Total System Memory	Bank B (left)				Bank A (right)			
	Socket 8	Socket 7	Socket 6	Socket 5	Socket 4	Socket 3	Socket 2	Socket 1
1 MB	-	-	-	-	256K	256K	256K	256K
2 MB	256K	256K	256K	256K	256K	256K	256K	256K
4 MB	-	-	-	-	1 MB	1 MB	1 MB	1 MB
5 MB	256K	256K	256K	256K	1 MB	1MB	1 MB	1 MB
8 MB	1 MB	1 MB	1 MB	1 MB	1 MB	1 MB	1 MB	1 MB

3

For modifier...	enter...
A through E	*the modifier* and continue entering the objective codes.
F	**F** and go to the next step.
G	**G** and go to Step 6.

FIGURE 11-8 Examples of Various Decision Tables

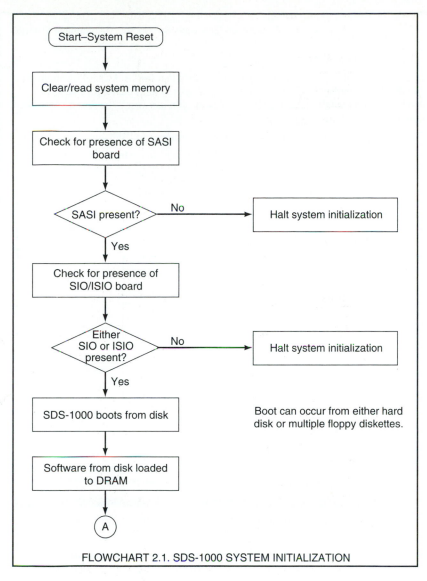

FLOWCHART 2.1. SDS-1000 SYSTEM INITIALIZATION

▬▬▬ **FIGURE 11-9** Flowchart

off in some fashion, either in a box, by a bulleted list, or by another graphic technique. Sometimes interesting related information can be placed in a "sidebar" column to the right or the left of the main text.

With the advent of the Macintosh, "*icons*" (visual symbols) have become essential for much of the software produced for personal computers today, and the writers who design the manuals use them in the hard-

ORACLE Database Gateways Set Up and Maintenance Tutorial

3.5.2 Verify a Schema

Under some circumstances, you may want to verify that the ORACLE information in an ORACLE gateway schema matches the data definition of the target ORACLE database. With the ORACLE Database Gateway, you can verify the schema information one file at a time.

The following steps show you how to verify the customer ORACLE gateway file.

1. Select **Database** from the Data Dictionary Main Menu. Type O for ORACLE Utilities, then type V for Verify ORACLE File Definition. The following screen appears:

```
PROGRESS  Data  Dictionary            Verify  ORACLE  File  Definition
Modify-Schema  SQL     Database Admin   Utilities   Reports  Exit

                    ┌─────────────────────────────┐
                    │ File name:                   │
                    │   CUSTOMER                   │
                    ├─────────────────────────────┤
                    │   AGEDAR                     │
                    │   CUSTOMER                   │
                    │   ITEM                       │
                    │   MONTHLY                    │
                    │   ORDER                      │
                    │   ORDER LINE                 │
                    │   SALESREP                   │
                    │   SHIPPING                   │
                    │   STATE                      │
                    │   SYSCONTROL                 │
                    └─────────────────────────────┘

   Database:  oracdemo (ORACLE)          File:
```

Figure 11-10 Screen Display (Used by permission of Progress Software.)

copy user instructions as well. These symbols decrease the number of words necessary on the page because they graphically represent certain tasks users need to perform or indicate possible user actions. A trash can, a notepad, a hand, and almost unlimited other possibilities can serve as helpful user cues throughout the documentation. In online doc, icons serve as "buttons" users can click on to do tasks (such as pulling down a menu or activating the delete function). In hardcopy, icons give the user a signal— for example, a pointing hand indicates that the information is important, or a question mark may indicate that there is an extra-help section for certain material. Figure 11-12 gives some examples of common icons, although these are just a few of the many that occur in today's software.

INPUT/OUTPUT EQUIVALENT SCHEMATICS

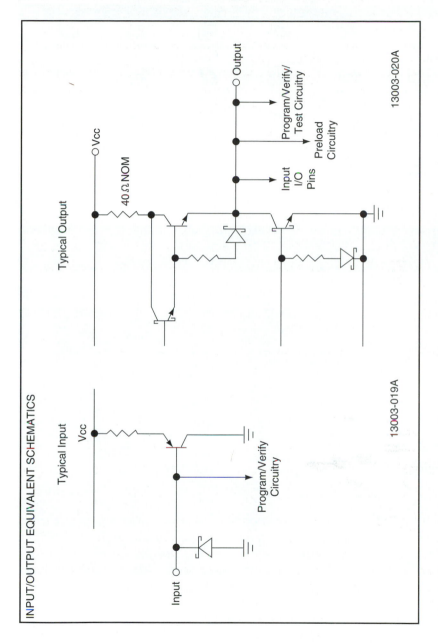

FIGURE 11-11 Schematics (Copyright © Advanced Micro Devices, Inc. 1992. Reprinted with permission of copyright owner.)

151

FIGURE 11-12 Common Icons

A word of caution about using these readability-enhancing aids: Be careful not to overuse them. It is easy for a novice page designer to think that adding more icons, rules, bullets, and so forth will make the page look more interesting. Actually, adding too many of these visuals may create confusion and overwhelm the user. Use them only if they make it easier for a reader to follow the text. Do not use them merely for decoration.

PLACEMENT OF GRAPHICS

One simple rule applies to graphic placement. Always place the graphics as close as possible to the text that describes them. But even this straightforward rule requires some corollaries. As a general principle, make sure to place the explanatory text first and the graphics second. Otherwise, users may not clearly understand what they are looking at when they are confronted with a picture before the prose. You might also follow the graphic with further textual explanation to be sure the user draws the appropriate conclusion. And in some instances, especially in technical reports and proposals, the graphics work best collected in an appendix in the back of the document so that they do not interrupt the prose information. In computer documentation, this "graphics in the back" arrangement is rarely effective.

To date, researchers have not been able to determine which is more effective: graphics on the left and text on the right, or text on the left and graphics on the right. Different designs work in different circumstances.

Reading habits have suggested, however, that visuals placed in along the fold (whether that means the left or right side of the page) are not as readable as those placed in the center or on the outside column of any page (See Figure 11-13).

The key to placing graphics effectively is to locate them so that the user does not have to flip pages between the text and the visuals, and so that they are large enough for the user to see them clearly.

CHOOSING A PAGE FORMAT

Deciding on a design for the page involves several preliminary steps. First, you need to determine the page size. Many software manuals are smaller than the 8 1/2 × 11 page. Is the smaller size right for your book? Second, you need to decide if your design will be modular—modular design requires a page format that can handle varying amounts of information while remaining consistent in format. Once you have made these decisions, your other decisions about page layout will follow more easily.

To design professional-looking computer documentation, think of the page as divided into a series of columns. Although the user cannot see actual vertical lines dividing the page, the text and graphics are organized cleanly within these columns. By having such a layout in mind, writers can produce documents that have a consistent format and appear as visually

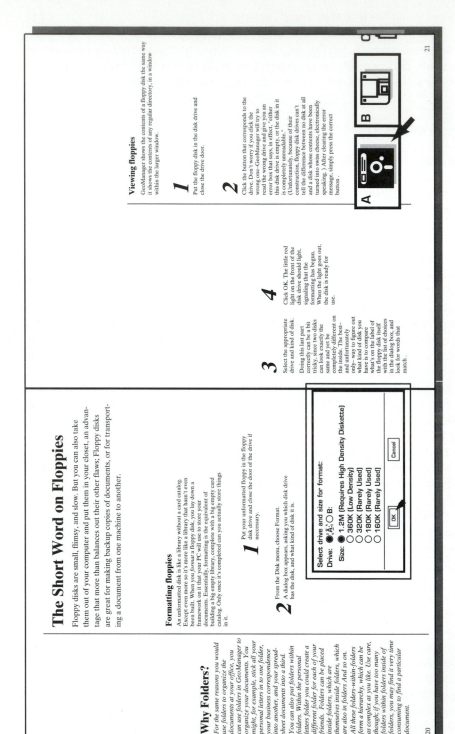

The Short Word on Floppies

Floppy disks are small, flimsy, and slow. But you can also take them out of your computer and put them in your closet, an advantage that more than balances out their other flaws; Floppy disks are great for making backup copies of documents, or for transporting a document from one machine to another.

Formatting floppies

An unformatted disk is like a library without a card catalog. Except even more so it's more like a library that hasn't even been built. When you *format* a floppy disk, you lay down a framework on it that your PC will use to store your documents. Essentially, formatting is the equivalent of building a big empty library, complete with a big empty card catalog. Only once it's completed can you actually store things in it.

1 Put your unformatted floppy in the floppy disk drive and close the door of the drive if necessary.

2 From the Disk menu, choose Format.

A dialog box appears, asking you which disk drive has the disk, and what kind of disk it is.

Select drive and size for format:

Drive: ● A: ○ B:

Size:
○ 1.2M (Requires High Density Diskette)
○ 360K (Low Density)
○ 320K (Rarely Used)
○ 180K (Rarely Used)
○ 160K (Rarely Used)

[OK]   [Cancel]

3 Select the appropriate drive and kind of disk.

Doing this last part correctly can be a bit tricky, since two disks can look exactly the same and yet be completely different on the inside. The best–and unfortunately only– way to figure out what kind of disk you have is to compare what's on the label of the floppy disk itself with the list of choices in the dialog box, and look for words that match .

4 Click OK. The little red light on the front of the disk drive should light, signaling that the formatting has begun. When the light goes out, the disk is ready for use.

Why Folders?

For the same reasons you would use folders to organize the documents at your office, you can use folders in GeoManager to organize your documents. You might, for example, stick all your personal letters in to one folder, your business correspondence into another, and your spread-sheet documents into a third.

You can also put folders within folders. Within the personal letters folder you could create a different folder for each of your friends. Folders can be placed inside folders, which are themselves inside folders, which are also in folders And so on.

All these folders-within-folders form a hierarchy, which can be as complex as you like. Use care, though; if you have too many folders within folders inside of folders, you may find it very time consuming to find a particular document.

Viewing floppies

GeoManager shows the contents of a floppy disk the same way it shows the contents of any regular directory, in a window within the larger window.

1 Put the floppy disk in the disk drive and close the drive door.

2 Click the button that corresponds to the drive. Don't worry if you click the wrong one–GeoManager will try to read the wrong drive and give you an error box that says, in effect, "either this disk drive is empty, or the disk in it is completely unreadable." (Unfortunately, because of their construction, floppy disk drives can't tell the difference between no disk at all and a disk whose contents have been turned into swiss cheese, electronically speaking.) After clearing the error message, simply press the correct button.

A   B

20   21

— **FIGURE 11-13**   Visuals Placed Incorrectly along the Fold (Used by permission of GeoWorks, Inc.)

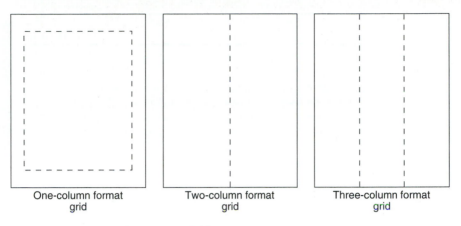

One-column format
grid

Two-column format
grid

Three-column format
grid

FIGURE 11-14 Column Format Grids

logical as they are verbally so. Formats that are most common in computer manuals are one-, two-, and three-column design (See Figure 11-14).

In a *one-column format,* as shown in Figure 11-15, the text takes up most of the page (as it does in this book), allowing only for top and bottom, right and left margins. This works if the book is small and the lines of text are not too wide. A very long line is hard for the eye to follow and slows reading speed. Another feature of this format is that graphics take up most of the page width and are generally at the top half of the page or at the bottom half, allowing the graphics to be large while still contained on the same page as the explanatory text. For this reason, documents including screen displays are often done in one-column formats.

A *two-column format,* shown in Figures 11-16 and 11-17, permits a number of options for text and illustration placement. In many cases, the text will run in one column, top to bottom, leaving the left or right column available for graphics. This layout is often the case with instruction sets. But a two-column format also allows you to create a type of checkerboard effect, also shown in Figure 11-17. By placing an imaginary grid over a two-column format, you can balance text and graphics in alternate columns. This is a fairly complex design, but once you know how to use a grid, you can manipulate your text and graphics as visual elements, placing them for maximum effect.

In a *three-column format,* the same principles apply as they do in two columns. You can move the text and graphics around on a grid, creating a professional-looking yet varied pattern. However, three-column layout is most often used by spreading the text over two columns and leaving the third column available for graphics and headings (see Figure 11-18). Only in rare instances does computer documentation actually appear in three narrow columns on the page. Such a design is too "scattered" for effective readability (see Figure 11-19.)

Top Output

Top output feeds each printed sheet face down into the bin at the top of the machine. Because it requires less space than rear output and stacks the sheets in the order printed, this is the output method most commonly used for standard types of paper. The bin can hold up to 200 sheets.

Rear Output

Fear output feeds paper straight through the printer and stacks the printed sheets face up (in reverse order) at the back of the printer. Because the paper path is flat, you should use this output method for specialty papers that resist bending, such as heavy stock, transparencies, and envelopes.

To use the rear output feature, pull down the stacker base and fold out the rear paper guides as shown:

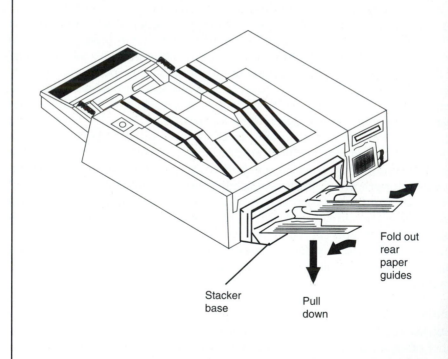

FIGURE 11-15   One-Column Format

FILLET CURVE

Filleting a Curve Parallel to a Plane

1 Enter FILL CURV PLAN, any other modifiers necessary, and a colon.

2 Digitize two or more curves and enter a semi-colon.

3 Enter **PLAN** to indicate you will digitize a plane. Digitize the plane and press ⬅

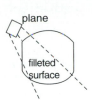

FILL CURV PLAN RAD .5 NEW TRIM:

ent *.11 .12.*

ent PLAN ⬅

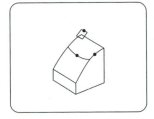

Explanation
A fillet with a radius value of .5 is created parallel to the plane indicated. The original curves remain and two new curves, trimmed back to the fillet, overlay them.

FIGURE 11-16 Two-Column Format

Speeding Public Access in Ireland
By Paul Carlson

This spring, Ireland's Companies Registration Office became the first Irish government agency to begin converting its paper files to computer images. A Wang Integrated Imaging System (WIIS), incorporating a VS 7310 mid-range computer, a 50-page-per-minute high-speed scanner, and 25 imaging workstations, is helping the CRO manage documents filed by more than 150,000 businesses registered in Ireland. Document images are stored on optical disks housed in a "jukebox" with capacity for up to 10 years' documents.

NATIONAL REGISTRY

An agency of Ireland's Dept. of Industry and Commerce, the CRO is the national repository of company documents required under Irish law, including incorporation documents and annual returns listing directors, shareholders, and financial data. CRO manager Paul Farrell compares his office to the Divisions of Corporation in some US states, except that the CRO maintains documents for companies throughout Ireland.

The CRO employs 58 people, 10 of them in document retrieval, the function being automated by the new WIIS system. Users of CRO data include lenders seeking company credit history, solicitors involved in buying corporate assets, and private citizens.

Installation of the WIIS system follows a pilot test aimed at evaluating Wang's imaging capabilities for the CRO. The development team, headed by Wang Ireland project manager David Murphy and senior systems analyst Alan Graham, installed a Wang VS 6E with optical disk drives linked to a duplicate CRO database to demonstrate potential gains in efficiency and improved public service. Overseeing the pilot test was a project

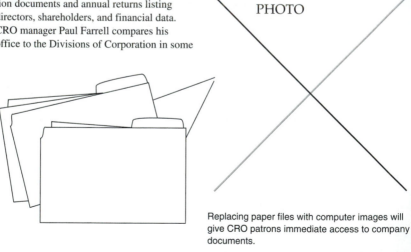

Replacing paper files with computer images will give CRO patrons immediate access to company documents.

FIGURE 11-17   Two-Column Format (Used by permission of Wang Laboratories, Inc.)

Data Limits

The next table lists data limits for line charts without markers, which can lower data limits.

Number of lines	Data points per line
1–4	500
5–8	225

Features

Special features for line charts include axis scaling, labels and values, and lines and markers.

Axis Scaling

Only the y-axis (vertical axis) is scaled. The scaled axis position may be left, right, both (same scale), or separate scales. Scaling may be linear or logarithmic.

Labels and Values

Data values determine the placement and shape of each line. You can use negative values in line charts. Type a label for each line and then use the Labels Style pop-up form to choose their color, font, and size.

NOTE Freelance Plus ignores blank values in line charts. Freelance Plus treats blank values in area charts as zeros.

Lines and Markers

Use the Line Style pop-up form to choose color, style, and width for each line.

You can mark the data points on each line. Use the Marker Style pop-up form to choose marker type, color, and size.

Style Tips

- For clarity, limit the number of lines to three or fewer. Create additional charts for additional comparisons.

- If your chart has more than one line, differentiate each line with a special color, line width, or line style.

- Use markers to emphasize data points. Omit markers to show a trend.

- Emphasize trends rather than individual values by smoothing the angles of lines in your chart. Add your chart to the current page with the Chart Go command. Then select and smooth the lines. See Chapter 6, "Go," and the section "Smooth" in Chapter 13, "Edit."

FIGURE 11-18   Three-Column Format

NEWS

UK User Conference Sets '90s Mark

The Wang UK User Group, an independent national group, hosted "Signposting the '90s," an attendance record-breaking annual conference and show, May 21–22 at the London Press Centre. In its April 30 newsletter, the group credited Wang President Rick Miller for keeping his commitment to send R&D experts and senior managers directly into the customer base by providing a slate of well-qualified and highly placed Wang speakers.

Speakers included Ken Olisa, vice president of Wang's Europe, Africa, and Middle East theater; Gerry Paul, vice president, communications systems; Steve Levine, director of core applications; Mike Runge, director of imaging development; Ken Osowski, director of UNIX systems product

management; and several other top corporate and UK Wang people.

Speakers from outside Wang included Bernard Smith, director, Artificial Intelligence Ltd.; Paul Carpenter, managing director, DMR Computer; Simon Holloway, director, Database Consultants Europe; and John Foster, information systems manager, Australian Treasury.

Topics ranged from Wang's Innovation on Standards strategy to the pros and cons of a distributed database, to imaging and Freestyle, to UNIX.

For more information on UK User Group activities, contact the Wang UK User Group, 60–63 Victoria Road, Surbiton, Surrey KT6 4NW; telephone 08 390 8824; facsimile 08 399 1360.

started up a special interest group under AUSWU for WIIS. For more information, contact Altit at 08 233 1608.

Chicago on the Move. The Chicagoland Wang Users Group has moved from 3227 N. Wilke Road, Suite 2706, Arlington Heights, IL 60004 to 3365 N. Arlington Heights Road, Suite J, Arlington Heights, IL 60004. Its telephone numbers, (voice) (708) 255-3003 and (facsimi-

le) (708) 577-7276, remain the same.

New User Group. The TriState Wang Users Group is a newly formed group for users from Indiana, Kentucky, and Ohio in the Cincinnati area. Meetings will be held on the second Wednesday of the month this

June, September, and December. For more information contact Jo Ann Glenn, TriState president, at Frost & Jacobs, 2500 Central Trust Center, Cincinnati, OH 45202; telephone (513) 651-6919.

Officer Elections. The VS Users Group of Central Pennsylvania reelected its 1989 officers slate to serve in 1990. The new officers are: Jim Lawrence, president; Rick Herr, secretary; and

---

FIGURE 11-19   "Scattered" Three-Column Format (Used by permission of Wang Laboratories, Inc.)

1. Find three computer manuals and identify the page formats. Are they done in one-, two-, or three-column layouts? Explain how you can tell.

2. Take one page of an existing computer manual and re-design it into a different page format. If it is one-column, change it to two- or three-column or vice versa. How does the new design change the readability of the information?

TYPOGRAPHY

With the many new software packages available these days, most writers have the technology not only to design page formats, but also to vary the type styles in their documents. It may help to have a quick guide to the basic typography terms.

A *typeface* is a style of type; for example, "Helvetica" or "Times Roman" (see Figure 11-20). A complete set of letters in one typeface, as well as all other type attributes such as boldface, italics, and so forth, is called a *font*. Most typefaces come in *serif*, meaning that the type has the extra flourishes on each letter, or in *sans serif*, meaning that the type has no such flourishes. For lengthy documents, serif type is more legible than sans serif, although sans serif works fine for short pieces and is preferable for headings and subheadings.

Type size is measured in *points*. The higher the point size, the larger the type. Generally, 10- to 12-point type is the most legible for computer documentation. Figure 11-21 shows the various type sizes.

DESKTOP PUBLISHING

Many companies, especially smaller ones, produce their documentation in-house by using desktop-publishing equipment. Writers who work in these environments are often responsible for writing as well as producing the manuals. Without specific training in design, writers are sometimes at a loss about how to use desktop-publishing technology to design effective documents. The following is a list of suggestions for desktop publishers[**]:

1. Avoid wide columns of type that extend from margin to margin. This long line length is very difficult to read.

[**]Much of this material comes from Roger C. Parker, "The Worst 25 Desktop Publishing Mistakes," *PC/COMPUTING*, May 1989.

American Typewriter Bold
American Typewriter Medium
Avant Garde Book
Avant Garde Book Oblique
Avant Garde Demi
Avant Garde Demi Oblique
Bookman Demi
Bookman Demi Italic
Bookman Light
Bookman Light Italic
Clearface Black
Clearface Black Italic
Clearface Bold
Clearface Bold Italic
Clearface Heavy
Clearface Heavy Italic
Clearface Regular
Clearface Italic
`Courier Bold`
`Courier Bold Italic`
`Courier Oblique`
`Courier Regular`
Eurostyle Bold
Eurostyle Bold Oblique
Eurostyle Demi
Eurostyle Demi Oblique
Eurostyle Oblique
Eurostyle Regular
Futura Book
Futura Bold Oblique
Futura Book
Futura Book Oblique
Futura Extra Bold
Futura Extra Bold Oblique
Futura Heavy
Futura Heavy Oblique
Futura Light
Futura Light Oblique
Futura oblique
Futura Regular
Garamond Three Bold
Garamond Three Bold Italic
Garamond Three Italic
Garamond Three Regular
Helvetica Neue Black
Helvetica Neue Black Italic
Helvetica Neue Bold
Helvetica Neue Bold Italic
Helvetica Neue Heavy
Helvetica Neue Heavy Italic
Helvetica Neue Italic
Helvetica Neue Light
Helvetica Neue Light Italic
Helvetica Neue Medium

Helvetica Neue Medium Italic
Helvetica Neue Roman
Eras Bold
Eras Book
Eras Demi
Eras Light
Eras Medium
Eras Ultra
Melior Bold
Melior Bold Italic
Melior Italic
Melior Regular
New Aster Black
New Aster Black Italic
New Aster Bold
New Aster Bold Italic
New Aster Italic
New Aster Regular
New Aster Semi Bold
New Aster Semi Bold Italic
New Baskerville Bold
New Baskerville Bold Italic
New Baskerville Italic
New Baskerville Roman
OCRA
OCRB
Optima Bold
Optima Bold Oblique
Optima Oblique
Optima Regular
Palatino Bold
Palatino Bold Italic
Palatino Italic
Palatino Roman
`Prestige Elite Bold`
`Prestige Elite Bold Slanted`
`Prestige Elite Regular`
`Prestige Elite Slanted`
Sabon Bold
Sabon Bold Italic
Sabon Italic
Sabon Roman
Times Ten Bold
Times Ten Bold Italic
Times Ten Italic
Times Ten Roman
Trump Mediaeval Bold
Trump Mediaeval Bold Italic
Trump Mediaeval Italic
Trump Mediaeval Roman
Univers Bold
Univers Bold Oblique
Univers Oblique
Univers Regular

FIGURE 11-20   Sample Typefaces

This is 5 Point
This is 6 Point
This is 7 Point
This is 8 Point
This is 9 Point
This is 10 Point
This is 11 Point
This is 12 Point
This is 13 Point
This is 14 Point
This is 15 Point
This is 16 Point
This is 17 Point
This is 18 Point
This is 19 Point
This is 20 Point
This is 21 Point
This is 22 Point
This is 23 Point
This is 24 Point
This is 25 Point

This is 26 Pt
This is 27 Pt
This is 28 Pt
This is 29 Pt
This is 30 Pt
This is 31 Pt
This is 32 P
This is 33 P
This is 34
This is 35

FIGURE 11-21 Type Sizes

2. Use white space judiciously. Avoid leaving too much white space between the columns or crowding copy onto claustrophobic pages.

3. Make sure the hierarchy of information is clearly visible on the page. This visual organization allows readers to focus on what's important.

4. Do not overuse borders or boxes. Borders around every page make each page a self-contained visual unit and thus hinder the reader's page-to-page transitions. Instead of using borders surrounding the page, use horizontal or vertical rules along one or more sides of a page. Single horizontal lines are particularly effective for isolating header or footer information from the body of the text. Use boxes only to set off text that logically should be isolated.

5. Use consistent typefaces. Avoid using so many different typefaces that the result is a circus effect.

6. Desktop publishing programs make certain assumptions about letter spacing. But these assumptions are based on "average" letter pairs, so wide, unsightly spaces can result between certain letter combinations such as an uppercase "W" and a lowercase "a." For type set in larger than 14 points, get in the habit of *kerning*—reducing the space between particular letter pairs that fit together, such as "T" and "i." Using appropriate *leading* fixes the same problem for spacing between lines of text.

7. Avoid excessive screening of backgrounds. Often, novice desktop publishers try to emphasize text by screening the background to shades of gray. This technique usually makes the text words harder to read. Use screening only when absolutely necessary.

USING A PROFESSIONAL DESIGNER

If you are not using desktop publishing or doing production in-house, you may need to hire a professional designer to produce your documents. In this situation, a few simple steps can alleviate problems that might otherwise occur.

- *Specify exactly what you want.* Do this in conversation and in writing. Use samples of old art if you have them, and draw quick mock-ups of what you have in mind. If you have particular typographic or layout ideas that you want implemented—including logos, fonts, sizes, and so on—write them down as part of the design specification you give to the designer.

- *Specify deadlines.* You should request a date when a draft of the design will be ready for you to review, and you should plan to have at least two reviews prior to final approval.

- *Arrange photo sessions, if necessary.* The designer may want to include photographs in the manual and will need to have access to the material to be photographed. Be sure to arrange a time for the photo shoot that is convenient for both the photographers and the company. You may need to get security clearance for this project and an escort for the photographer.

- *Hire a professional illustrator.* If you want illustrations, hire a professional. Ask to see the illustrator's portfolio and check to make sure he or she has expertise in the specific technical area you need. As stated earlier, tell the illustrator exactly what you want in terms of size, detail, and perspective.

▬▬FINAL THOUGHTS

Effective graphics and page design take a great deal of care and expertise. As this chapter has shown, writers cannot automatically assume they are good designers simply because they are good writers. Nonetheless, readability in computer documentation depends on more than clear prose. It also depends on a clarity of design that gives users quick access to the information and allows them to find their way around in the document as viewers first, then as readers.

▬▬ EXERCISE ▬▬▬▬▬▬▬▬▬▬▬▬▬▬▬▬▬

A good way to test your use of these design tools is to try the "squint test." Take a copy of the page you are working on and tack it to a wall just far enough away so that you can see the page but cannot read the words on it. Is the hierarchy of information visually apparent? Can you see which material is important? Does the page design direct your eyes to the appropriate places, or is there too much visual clutter on the page for you to focus on what is important?

Try the squint test on the following pages (Figures 11-22 through 11-25) to determine if the page design works. Write a brief critique of each design, explaining why you think it works or not.

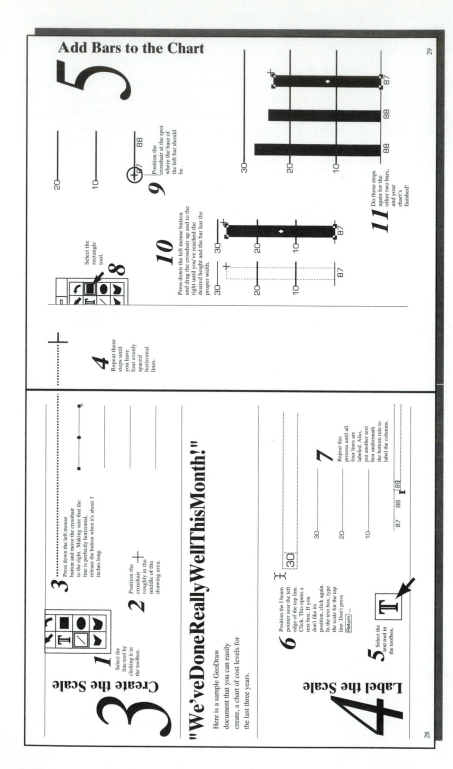

FIGURE 11-22 Sample Page 1 (Used by permission of GeoWorks.)

BASIC

VAX BASIC is the version of the BASIC programming language that is installed on the VAX computer. You can access VAX BASIC in two ways:

- as an ordinary compiler
- as an interactive system, the BASIC environment

All BASIC source files used with the ordinary compiler should have .bas as the filename extension. For example, the filename for a BASIC program might be *prog1.bas* where "prog1" is the filename and ".bas" is the filename extension.

Access as a compiler

The BASIC compile command takes the format:

basic *filename*

where *filename* is a BASIC program you have created and saved.

Options

You can add options, called *qualifiers* on the VAX, to the compile command using the format:

basic/*qualifier1/qualifier2/... filename*

Useful qualifiers for BASIC include:

/debug	Uses VAX/VMS Debugger and Run-Time Error Traceback
/list=*filename*	Produces *filename* containing source code listing with line numbers and error messages

For example, the command *basic/debug/list=myfile prog1* will compile the file *prog1*, debug it, and place the source code listing and any error messages in the file *myfile*.

After compiling, you can link and run your program using the Link and Run commands discussed on page—.

Getting Help

To obtain help with the BASIC compile command, type:

help basic

Documentation

Documentation for VAX BASIC is available (for reference use only) at the CRC office.

FIGURE 11-23 Sample Page 2

Batch Printing Chart and Lab Reports

The 780 RDM allows you to print chart and laboratory reports continuously or to stop printing between tests (for example, so that you can change the paper on which reports are printed). The batch print function also has a filing option. When you select this option, the 780 RDM files the reports after they are printed. Once reports are filed, they no longer appear on edit, print, and file lists.

> ■ NOTE   The printer must be turned on and online before you can print reports.

1. Turn on the printer and verify that it is online.

2. Select **ACCESS PATIENT INFORMATION** from the Main Menu. Press *ENTER*.

    The Access Patient Information menu is displayed.

3. Select **PRINT REPORTS** from the Access Patient Information menu. Press *ENTER*.

    The Select Reports to Print screen is displayed as shown in Figure 4–7. This screen lists the tests and types of reports that you can print.

FIGURE 4–7. Select Reports to Print Screen

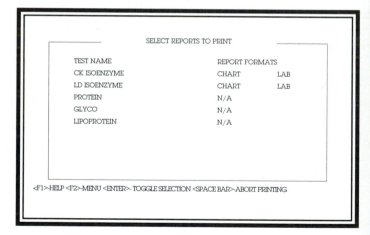

```
                         SELECT REPORTS TO PRINT

         TEST NAME                      REPORT FORMATS
         CK ISOENZYME                   CHART        LAB
         LD ISOENZYME                   CHART        LAB
         PROTEIN                        N/A
         GLYCO                          N/A
         LIPOPROTEIN                    N/A




  <F1>-HELP <F2>-MENU <ENTER>- TOGGLE SELECTION <SPACE BAR>-ABORT PRINTING
```

FIGURE 11-24 Sample Page 3

1   Getting Started

Procedures:

NOTE: Throughout the User's Guide, the CD-ROM drive is referred to as drive L:.
Your system configuration may specify a different drive as the CD-ROM drive.

1.   Make sure the MS-DOS CD ROM Extensions (MSCDEX) are running by typ-
     ing MSCDEX and pressing <RETURN>. The message MSCDEX Already
     Started" will verify this.

2.   Type L:\INSTALL and press <RETURN>. The following screen appears:

NOTE: Menu selections can be made by pointing or by the first letter of the
selection throughout all but the Database Installation menu.

3.   Select TEK Financial. The following screen appears:

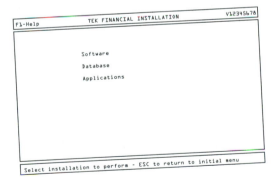

FIGURE 11-25 Sample Page 4

CHAPTER 12

TESTING
THE DOCUMENT

Many writers think their documentation projects are finished once they have completed writing, editing, and designing the material. They are mistaken. Before documents are ready to ship to the customers, they must undergo usability tests to determine how well they will work for users. These usability tests are an essential part of the documentation process and should not be omitted.

THE NEED FOR TESTING

Would you buy a car without test driving it? Probably not. In fact, most car buyers do more than just test-drive the product before they buy it, they also check *Consumer Reports* and *Road and Track* magazines to get professional opinions on its quality. Only after checking out the car thoroughly do smart buyers part with their money and drive off the lot in their shiny new purchase.

People interested in buying computer hardware and software do much the same thing: They test the products at the dealers' showroom or have marketing representatives provide demonstrations, then they read the product reviews available in most computer magazines. Not until they are satisfied that the product works well do they buy it.

But consumers rarely have the chance to "test- drive" the documentation that accompanies their new product. Instead, they have to trust that the manuals and online help will be easy to use and will quickly teach them to use or assemble their purchase. Imagine the frustration these users feel when the documentation proves to be more complex than the machine itself, or when the stack of manuals necessary to explain the software is more intimidating than helpful. Although documentation may be technically accurate and packaged in slick designs, it fails if customers cannot easily use it. Remember, users do not want to spend time reading documentation; they want to use the computer product. Anything that impedes their ability to perform the necessary tasks is a liability.

The only way to ensure that the documentation will work is for the writers to test it with real users before the final version is shipped. Even at that late date, there is still time to correct any usability problem that the tests reveal. Then, when the manuals are shipped with the product, both the writers and the customers can have confidence that the documentation is of high quality and will function well.

═══TYPES OF USABILITY TESTS

As in all phases of the documentation process, usability testing involves preplanning and careful implementation. The first step in designing an effective test is to decide specifically what you want to test. In other words, what qualities, variables, and characteristics are you testing, and what type of test will work best? Further, what kind of test can you afford?

There are four main types of usability tests: the written test, the task-oriented test, the attitudinal questionnaire, and the informal observation and interview. Each one measures different things.

- *Written tests* test objectives that do not involve procedural tasks. For example, if you want to test users' *understanding of concepts,* a written test is appropriate. You can use any of the traditional types of written examinations: multiple choice, matching, completion, or true-false.
- *Task-oriented tests* determine how well users can follow the documentation to complete specific procedural tasks. This type of test requires subjects to use the product in a simulated environment such as a test lab.
- *Attitudinal questionnaires* determine how the users feel about the documentation. These questionnaires consist of three parts that allow users to rate various aspects of a manual: (1) a series of statements about the usability of specific parts of the document, (2) a range of possible

responses to each statement, and (3) a comment section for each statement. See Figure 12-1 for a sample attitudinal questionnaire.

- *Informal observations and interviews* yield more qualitative than quantitative results. By watching people use the documentation and by talking to them about it, you may gain perspectives not provided by the other test types. However, this kind of test should be done in conjunc-

---

Directions: This questionnaire gives you the opportunity to share your opinion of the publication you have just reviewed. For each of the following statements, please indicate the response *closest* to your opinion by checking the appropriate blank. Please check only *one* blank for each statement. If you feel that you were not given sufficient opportunity to make a judgment, check the blank under the heading *Unable to Judge*.

If you check a blank that indicates a usability problem or check the blank under the heading *Unable to Judge*, explain why in the comments section for that statement.

		Agree	Neutral	Disagree	Unable to Judge
1	The objectives were clearly stated in the front matter (Preface or introductory section).	___	___	___	___
	Comments:				
2	The material was organized in a logical sequence.	___	___	___	___
	Comments:				
3	The headings used in the text made it easy to find information.	___	___	___	___
	Comments:				
4	The procedures needed to perform major tasks were adequately explained.	___	___	___	___
	Comments:				
5	Clear, relevant examples were used to illustrate the material.	___	___	___	___
	Comments:				
6	Most terms were adequately defined.	___	___	___	___
	Comments:				
7	Some essential topics were not included in the index.	___	___	___	___
	Comments:				

FIGURE 12-1   Attitudinal Questionnaire (From *Technical Communication,* Fourth Quarter, 1980. Used by permission of the Society for Technical Communication.)

tion with one of the other types. Qualitative information alone is not sufficient to evaluate the usability of a document.

━━DESIGNING A TEST

Designing an effective test, then, requires you to first determine your objectives (what you want to measure). You may decide to create a test that combines some of the standard types, or you may want to perform several different tests on the same document. The steps involved in designing a combination test are as follows:

1. *Select tasks and determine optimum time lengths.* What procedural tasks do you want the test subjects to perform during the test? How long should it take them to complete each one?

2. *Write a script.* You will need two scripts for the test: one for the test subjects and one for the monitors. The script the subjects see gives them the scenario in which they are to perform the tasks: the environment, the goals, the limitations. The script for the test monitors contains the subjects' script as well as any special instructions for administering the test.

3. *Find test subjects.* Try to find a sufficient number of participants to make the test's results worthwhile. If only a few people can participate, you will learn a little but not enough to make substantial changes in the document. You may use internal subjects from different departments within your company and external subjects from outside. Of course, the test will yield the best results if the subjects are actual customers for the product. A lot of customers *like* to do these tests. It gives them a sense of participation in the product design. In order to do testing with "real" customers, you must work with the sales and marketing staff to identify qualified subjects.

4. *Develop a pretest questionnaire.* This questionnaire is sent to the subjects before they take the test. Its primary goal is get background information about the subjects so that you can better analyze their performance. You may ask about familiarity with the product or similar products, computer experience, work environments, education level, and so forth. Again, the sales, marketing, and customer support people must be involved here.

5. *Arrange the testing place.* If your company has a human factors lab, use it. If not, try to find an empty training room or conference room where you can set up the equipment and remain undisturbed during the test period. Too many distractions for the participants will skew the test results.

6. *Test the test.* Do at least one practice run through the test with trial subjects from your own department or from elsewhere in the compa-

ny. By so doing, you may discover problems you can correct before administering the test to actual subjects. Check especially the placement of items on the test. The arrangement of parts can directly affect the results.

7. *Administer the test.*

8. *Conduct posttest interviews and/or develop a post-test questionnaire.* When you talk to the participants after the test, or when you receive their post-test questionnaires, you can gain valuable reactions to the product and to the test itself. Often the strict nature of the tests do not allow users to respond fully to the product, but they can fill in these gaps afterwards.

9. *Write follow-up thank-you notes.*

10. *Tally and analyze the test results.* As in the Needs Assessment described in Chapter 3, the analysis of this test works best when presented graphically as well as verbally. If the results are presented clearly, you can learn a great deal about the usability of your documentation, and you have sufficient evidence to support any changes you decide to make in it before the final version is ready to ship.

11. *Consider sharing a summary of the results with the testers.* Doing so closes the loop and makes them feel their efforts have been worthwhile.

No matter whether you design a combination test or use one of the standard usability tests, your documents will be better for having tested them. Writers who are committed to a documentation project and are working feverishly to meet their deadlines lose their objectivity about the material. Testing the document brings fresh eyes to the project and provides a kind of "reality check" for writers so absorbed in their work. Only after a document has passed these tests should it reach the customers' hands.

═══ EXERCISES ═══════════════════════════════

1. As a practice exercise, form a group of three or four people and design an attitudinal questionnaire about a computer product with which you are familiar. Once you have finished the questionnaire, send it to at least five test subjects who also have used the product. Write an analysis of the results. Afterwards, write a brief analysis of the questionnaire. What worked? What did not? How would you design it differently the next time?

2. Design a combination test following the steps listed in this chapter. You may have to modify the test somewhat to fit your circumstances, but try to keep it as close as possible to the design recommended here. After you administer the test, analyze the results and the test itself. What worked? What did not? How would you design it differently the next time?

CHAPTER 13

PREPARING AN INDEX

The most useful feature of hardcopy documentation is the index. Even with a table of contents, a manual without an index becomes frustrating for users who need to find something in a hurry. Yet with deadlines and budget cuts bearing down on documentation departments, some technical writers decide to cut the index to save time and money. Such a decision almost always cuts the usability of the manual by at least half. It's not a fair trade.

IS AN INDEX NECESSARY?

An index is more than an alphabetized list of key terms with page numbers; it is an information retrieval tool. As such, it functions as a map that allows users to navigate documentation in ways that suit their special needs. In one sense, an index is really a precursor to hypertext technology because it gives the users power to use information in any order they like. If you want to explore particular aspects of the product—or refresh your memory about a certain function—a good index allows you to override the

Table of Contents

FIGURE 13-1   Table of Contents That Is a List of Commands

table of contents to find what you need without plowing through the whole manual. For example, if you want to know how to rename a file, you don't have to read the manual from beginning to end simply to find that information. Instead, you check the index and find "Files, renaming, 2- 12." Quickly flipping to page 2-12, you find all you need to know about changing a filename. Indexes are an indispensable reference tool.

Nonetheless, there are some exceptions. In extremely short documents (under ten pages) or in reference guides that are arranged by a list of alphabetized commands or functions, indexes can be superfluous. In these instances, the table of contents becomes the index (see Figure 13-1).

ELEMENTS OF A GOOD INDEX

Good indexes are created because writers think ahead and plan effective strategies for the index components. The following are the basic components you have to work with as you develop your indexing strategies.

Key Terms: Important words and phrases that represent significant information the users might want to locate quickly. Figure 13-2 provides some help with determining key terms.

Main Entries: Arranged in alphabetical order, specifying document topics.

> *Documentation,* 2-1 to 2-10

Subentries: Specify particular aspects of the main entry.

> *Documentation,* 2-1 to 2-10
> editing, 2-3, 3-11

Levels: Each level of subentries in an index makes the information more specific. For example, an index with only one level of information is more abstract than an index with two levels. Figure 13-3 shows three examples of indexes at different levels of abstraction.

Cross References: Direct users to alternative entries or additional information.

SEE follows a main entry that has no page references:

> Manual. See *Documentation; Writing*

SEE ALSO follows an entry that has page references:

> Documentation, 2-1 to 2-10. See also *Writing*

Page References: Locate information about the entry in the document.

> Documentation, 2-1 to 2-10
> Editing, 23, 31-37

Synonyms: Alternative entries especially for new users who may not be familiar with the product's technical terms. By including synonyms to entries, you can capture terms that these readers might use.

Definitions: Include "defined" or "definition of" as a second- or third-level entry for key terms.

Effective indexes make the product easier to learn and use, especially when they contain multiple entries and cross references. If you follow a few simple rules, your index will work well.

- All main entries should be nouns or gerunds ("Formatting" or "Data Entry" instead of "How to Format" or "How to Enter Data").
- Two to three possible references under each main topic indicate a need for a subentry.
- The main entry should not be followed by a group of sub-entries that all refer to the same pages. Turn such sub-entries into main entries.
- Finish writing the document before you begin indexing it. Both jobs go faster when they are kept separate. (But the writer should remem-

Terms To Index	Typical Locations of Terms To Be Indexed
Key terms	In titles of sections, subsections, figures, and tables; in introductions to sections, subsections, and lists; and in the glossary
Synonyms and cross-references to each key term	Usually derived from key terms themselves, not within any particular location of the document
Key terms that identify design or analysis processes	In titles of sections, subsections, figures, and tables; in introductions to sections, subsections, and lists; in the glossary; and in first sentences of paragraphs
Key terms of tasks	In titles of sections, subsections, figures, and tables; in introductions to sections, subsections, and lists; in the glossary; and in first sentences of paragraphs
Key terms that identify structural characteristics of products, including hardware and software products	In titles of sections, subsections, figures, and tables; in introductions to sections, subsections, and lists; in the glossary; and in first sentences of paragraphs
Terms that are defined within the text	In introductions to sections and subsections; and in the glossary
Proper names	Throughout the text
Names of commands and functions	Throughout the text

━━ **FIGURE 13-2**   What Subject Matter to Index (From Sandra Oster, "Features of Indexes in Computer Documentation," *Technical Communication,* used by permission.)

ber throughout the writing process to use task-oriented headings and terms that will work well later in the index.)

- Think more of the user than the product.
- Be consistent in your index format.

Because most printed manuals require an index, technical writers should use these strategies to construct one that works well. "That's easy," you say, "I'll just use the new indexing software and generate an index in

One-Level Index	Two-Level Index	Three-Level Index

Falcon Framework
File system objects
Firmware
	Falcon Framework
	  defined
	  major components of
	File system objects
	  copying
	  defined
	  directory links
	  file links
	  moving
	  opening
	  unresolved links
	Firmware
	  debugging
	  defined
	  testing
		Falcon Framework
		  defined
		  major components of
		    AMPLE
		    BOLD Browser
		    Common User Interface
		    Data models
		    Decision Support System
		    Design Manager
		    Multi-window features
		File system objects
		  copying
		  defined
		  links
		    directory
		    file
		    unresolved
		  moving
		  opening
		Firmware
		  debugging
		  defined
		  testing

FIGURE 13-3   Levels of Indexing (From Sandra Oster, "Features of Indexes in Computer Documentation," *Technical Communication,* used by permission.)

no time at all." Not so fast. Before trusting a computer to do the work for you, think about your options.

GENERATING THE INDEX BY HAND OR BY COMPUTER?

Indexing by hand is a time-consuming, tedious process. To do it right requires going through every page of the document carefully, highlighting key terms, and creating an index card for each term. At the top of the card, you should write the key term and then list its various forms, synonyms, other words related to it, and the page numbers on which these items appear. Then alphabetize the cards and make sure all the key terms are cross- referenced. Once you have completed cards for all of the terms, you must type the information in a format that is accessible to the user.

In the age of technology, this process seems unnecessarily complicated. It's no wonder many writers turn to indexing software rather than depend on the handcrafted index. But even the most sophisticated software has limitations that require you to do some of the work by hand.

Indexing software comes in three types: Key word, embedded, and free-standing.

Key Word: An indexing software that comes with word- processing packages that functions primarily to list page numbers of words the user highlights in the text. Its shortcoming is that it does not discriminate between important and less important uses of a word; it searches the text and lists every page on which the word appears.

Embedded: Contained in desktop publishing software, this type of indexing creates an alphabetized index file that becomes very hard to edit. The user works with "tokens" provided in the software to produce the original index file. Unfortunately, any editing done later in the index file is not automatically incorporated into the embedded indexing tokens. These changes must be manually entered into the tokens.

Free-standing: Designed specifically for indexing from hardcopy, free-standing software creates an index for a hardcopy document such as a book or a manual that is already completed. Unlike the embedded indexers, these free-standing programs make editing the index easy, but the files they create are not linked to the document page numbers. The linking process must be done manually after the index has been generated.

As you can see, indexing software is useful, but it has serious shortcomings. Even if you choose to use one of these indexing types, you will have to do a significant amount of work by hand anyway, especially if your index is *user-oriented* rather then *product-oriented.*

▬USER-ORIENTED INDEXES

A good index is a guide for the user and therefore is designed to focus on the user's needs. When you plan an index, think more of the user than of the product. For example, the user-oriented indexer focuses on what users want to know and what terms they will have in mind. On the other hand, a product- oriented indexer thinks only about what information is on each page and what features are being described. As a result, the product-oriented index is often just an alphabetical list of commands and chapter headings. Only users who are familiar with the terminology of the manual can benefit from it.

Take the following example, developed by a professional indexer.[**] This excerpt from a manual would be indexed differently by user-oriented and product-oriented indexers. Note the different entries each indexer uses to list the information given in this excerpt:

[**]Susan Holbert, Indexing Services, Waltham, MA.

OUTLINING

The first step in developing an index (even if you are using indexing software) is to work by hand. Print out a copy of the document and work with the hardcopy. As you review the material, underline any words that you think should be main entries. During this same process, specify any sub-entries that become apparent and mark important cross references. It may help to use different colored pens to complete this stage—one color for the main entries, one for the sub-entries, etc. You will also need to have separate sheets of paper (or index cards as mentioned earlier in this chapter) on which to take notes and organize the jottings and underlinings you have made in the document. You may also want to think about synonyms for these entries at this point, while you have the context in front of you.

DRAFTING

The first draft stage is when you collect all of the notes you took in step one and make tentative decisions about organization and structure. List the main entries in alphabetical order and include sub-entries under them. Add page numbers and cross references.

REVISING

This stage is the main part of the indexing process and requires the majority of your time. The tasks you need to complete during this phase of indexing include refining and simplifying entries, clarifying overlapping entries, combining or separating entries, deleting references to identical material, adding more cross references, adding page numbers, making sure you have used the right level of generality, checking for consistent terminology and format, and so on.

TESTING

The index should be tested to assure that it works well, especially if changes in the document are made at the last minute and the index is affected. In any case, at least one other pair of eyes should review the index for its usability and completeness. Someone who represents the real audience for the document is the best tester. That person will not know the product well enough to get along with a product-centered index.

When you have finished all three indexing stages, you should have a usable, concise index that will serve the users well. Remember, effective indexing takes patience, time, and planning. Anything less produces an index that may waste users' valuable time by providing unintelligible maps, sending them down dead ends, or frustrating them with lengthy detours.

Moving between Applications: To use a WORDSOUP file in another application, save it in nondocument mode.

Product-oriented index	*User-oriented index*
SAVING: in nondocument mode	Exporting Files
	Importing Files
	Transferring Files
	FILES: exporting
	importing
	transferring
	APPLICATIONS: moving between
	Nondocument Mode

The user-oriented indexer tries to generate as many alternative entries as possible and uses synonyms for the key terms to increase the users' chances of finding information. The product-oriented indexer depends on terms related to the product, terms users may not know unless they are already familiar with the document. To create an effective user-oriented index, you should realize that the index will require extensive editing. Whether or not you choose to use indexing software, you will need to write a first draft and then revise it over and over to make it genuinely helpful for the user. Such editing requires time and a great deal of manual work.

=== EXERCISES ===

Think carefully about possible index entries for the following topics:
Creating New Files with Open Command
Utility Macros
Searching for and Replacing Text
Saving Text Files as ASCII Files
Make a list of the possible entries and determine which are user-oriented and which are product-oriented.

THE INDEXING PROCESS

Developing an index does not begin with the writer sitting down to type in entries. Instead, the indexing process resembles the writing process in its stages of outlining, drafting, revising, and testing.

The following page from a computer manual contains enough information to serve as good practice for indexers. Develop a list of alphabetized index entries for this page. Remember to use multiple alternative entries and sub-entries where appropriate. In completing this exercise, follow the indexing process discussed in this chapter.

Moving Text

When you move text, you delete it from its original location and insert it at a new location.

To move text with keys

① Select the text you want to move.
② To delete the text from the document and place it in the scrap, choose Edit Cut (ALT, E, T).
③ Move the cursor to where you want to insert the text.
④ To insert the contents of the scrap at the new location, choose Edit Paste (ALT, E, P).

To speed move text with the mouse

You can move text quickly with the mouse without first moving the text to the scrap.
① Select the text you want to move.
② Point to where you want to move the text.
③ Hold down CTRL and click the right mouse button.

Macro Hint You can speed the task of moving text by using the supplied macro *move_text*. For more information, see Chapter 22, "Macros."

Editing Shortcuts

To delete, insert, move, and copy text quickly, use the following shortcut keys:

To	Press
Delete text to the scrap	SHIFT+DEL
Insert text from the scrap	SHIFT+INS
Copy text to the scrap	CTRL+INS
Delete text permanently	DEL

Note If you want the INS and DEL keys to work as they did in Word, version 5.0, choose Utilities Customize and turn off the Use INS For Overtype Key check box.

CHAPTER 14

ONLINE
DOCUMENTATION

In a world nearly overwhelmed by the volume of paper produced daily, online documentation is a welcome respite. And, it makes sense for the computer industry to use its main tool—the computer—as a means of instruction instead of adding unnecessarily to the paper glut. Online documentation makes it easier for users to get the help they need without having to leave the screen, and it has opened exciting new possibilities for communicating information. Dialogue boxes, help screens, pull-down menus, and other online tools make information almost immediately accessible to users. With the advent of hypertext technologies, users can now take much more control of their learning processes and often navigate through information in the order and at the speed they choose.

But online documentation cannot do everything. It has its place and purpose, just as hard-copy manuals have theirs. Any discussion of online documentation should first begin by defining the differences between the two media.

▄▄▄THE DIFFERENCES BETWEEN HARD COPY AND ONLINE DOCUMENTATION

Online documentation differs from hardcopy in many important ways, but especially in what it is for and how it is used.

WHAT IT IS FOR

Unlike hardcopy manuals, online documentation is not a complete user guide or reference manual. It does not have all the information neatly packaged in a series of chapters that cover everything about the product. Instead, most online documentation is similar to quick reference material available only when users need it. For more complete information, users must return to the offline manuals. The two primary purposes for online material are:

1. to give users quick instructions about how to use the product, and
2. to serve as a navigational tool through the software.

Hardcopy documentation, on the other hand, provides complete information about the product and enables users to both understand and perform the procedures. Hardcopy manuals also show the relationship of all the pieces of information because the user can physically see all the different sections of the manuals, can realize how long each section is and in what order it comes, and can understand from the table of contents how everything fits together into a body of coherent material. In contrast, online documentation isolates information into totally separate units because of the limitations of the screen. When users want information, they need to know in advance exactly what kind of material they need to call up on the screen, and the documentation provides only that circumscribed material. To learn more, users must call up different screens. This "layering" of information screens sometimes makes it difficult for users to grasp the overall coherence of the material. And the process of moving through more and more layers of on-screen information makes it especially difficult to retrace steps to read something again quickly. In hardcopy manuals, it's easy to flip back and forth among the pages, and even to look at two pages simultaneously.

HOW IT IS USED

Users read online documentation much differently from the way they read hardcopy manuals. Manuals are written in a linear fashion—that means they have a set beginning, middle, and end. When users open the book, they follow the reading path set up by the writer, even though they may jump around a bit, especially when they use the manual for reference. Nonetheless, hardcopy manuals require users to read in a careful, linear way. They also require users to leave the computer screen to page through the material. Anyone who has had to juggle a large manual on a small computer table (a table already cluttered with a keyboard, a computer, a printer, and probably a coffee cup) understands how frustrating reading a manual can be.

Online documentation, on the other hand, appears directly on the screen whenever users need it. If you want to know about type fonts, you can request a list of them by pressing a function key or clicking on an icon conveniently located on the screen. If you need help remembering how to finish a procedure, call up a help screen and then return to your original work. This convenience and immediacy are two of the most attractive features of online documentation.

Even more convenient for some applications are hypertext tools. These tools let users "read" the material in various ways. For example, in some hypertext applications, you can call up a screen that tells you about a particular topic's background or call up one that explains how that topic is linked to other subjects. If that's not enough, you can ask for definitions of terms or pictures of the subject. Hypertext programs are like large databases of information that allow users to follow various paths to retrieve the material stored within the program. The more advanced the program, the less linear it is. In other words, users don't have to follow a single straight path to retrieve the information; they can take a variety of shortcuts and side trips to reach the same destination.

But online documentation has drawbacks, too. The fact that it is often nonlinear and encourages different patterns of reading creates problems for writers who are used to composing hardcopy manuals. Hypertext technology, in particular, calls for a different rhetoric—not that of linear, book progressions. Writers need to learn how to sustain a text without using "propositional" logic—that is, without using logic that depends on a stipulated step-by-step pattern. Readers see only one screen at a time, and there is no guarantee how they will proceed from there.

Furthermore, online documentation can only be presented one screen at a time. That makes it difficult for users to flip back and forth through long passages of text or complex material. It is also more difficult to read text on screen than to read it on paper because of eye strain. As a result, technical writers have done a lot of research on how much information can comfortably appear on the screen and how the online material should be designed for ease of readability. Large descriptions or long explanations do not work well online, nor do instructions that have a lot of parts. For these kinds of information, users prefer hard copy.

In short, writers cannot design successful online documentation by using the same assumptions they use for paper manuals. Figure 14-1 provides a summary of the basic differences between online and paper media. Each kind of documentation has benefits and each has liabilities. Writers should use the medium that best fits their situations, and they should design the documentation *from the beginning* for paper or screen.

Feature	Paper	Online	Implications for online
Display area	A two-page spread or at least a full page, typically 600 sq cm.	The window or screen typically displays 1/3 to 1/6 as much as a single paper page.	Reading passages must be much shorter. Relationships implied by positioning on the page are lost.
Display shape	Taller than wide (primarily for binding strength).	Wider than tall (to match human field of vision).	Vertical page layouts require excessive scrolling.
Resolution	10,000 dots per sq mm for typeset documents.	10 dots per sq mm for "high" resolution screen.	Less fine detail and fewer thin lines must be used. Typefaces may have to be larger.
Physical presence	Books have a tangible presence; position is clearly indicated by the size of stack of pages to left and right.	Online documents have no physical presence.	Users do not know how much information is available. Users lack simple navigational references. Users are not intimidated by large documents.
Dynamics	Books are static.	Computer displays can be dynamic and interactive.	Online documents can use animation to communicate concepts involving change and motion.
Other media	Books are limited to text and static graphics. Color is expensive.	Increasingly, computers can include color, sound, music, motion—just like TV.	Users rapidly lose interest in static displays.

FIGURE 14–1 Differences between Paper and Online Documentation (From William Horton, "The Wired Word," *Technical Communication,* Second Quarter, 1992. Used by permission of Society for Technical Communication.)

Find a hardcopy computer manual and read through it quickly. What parts of it could benefit from an online supplement? Why would online copy be helpful to the user? How would the online material differ from the hard copy so that it would be especially efficient for the user? Explain in detail your answers to each of these questions.

══TYPES OF ONLINE DOCUMENTATION

It is a common mistake among computer users to think all online documentation is the same. Nothing could be farther from the truth. Even though all the various types share some commonalities (they are presented on screen, they are made up of short pieces, they are user-controlled, and so on), they each have separate functions and therefore separate design considerations. This section discusses some of the more familiar online documentation types, although many more exist and are being invented everyday.

HELP SCREENS

Help screens are embedded in most software programs and provide users an easy way to get help without turning to the printed manual. For example, users who are using a word-processing program and need a quick refresher on how to use the commands can press the appropriate help key and a list of the commands appears on the screen. In many cases, the "help" material is visible in a window that pops up in the middle of the original screen—thereby allowing users to remain in the context of their work while still viewing the help. Sometimes, however, the help screen is a totally new screen that replaces the original work, although that significantly interrupts the reader's train of thought and is less desirable. In either case, pressing the "Escape" key removes the help. In a few short seconds, users can access help, receive the information they need, and return to work.

As writers assigned to design help screens, remember to keep the information as concise as possible (lists are especially helpful formats to use) and to attempt to contain all the necessary material on one screen. If users have to scroll to multiple screens, the help becomes more cumbersome than it should be. Another important element to remember about help screens is that you should make it clear on each screen how the users can leave the help and return to their original work. The directions for escaping help should appear in the same place on every help screen. Figure 14-2 shows an effective help screen.

```
CM
PRMPT: PRESS ANY KEY TO REMOVE HELP
INDEX WORD KEY ASCII ERROR HELP COMMANDS FUNCTIONS TEMPLATE DOS
```

INDEX TO XYWRITE

This HELP file provides you with an alphabetical index of all of the topics covered by HELP. Type the first letter of the topic that you are interested in. With each topic there is a phrase in bold. The bold represents one of 4 things;

1. a descriptive phrase
2. the command name
3. the key combination that activates that function
4. the 2 character function call for keyboard commands

Put your cursor on the topic that you want and hit return.

```
    A   B   C   D   E   F   G   H   I   J   K   L   M
    N   O   P   Q   R   S   T   U   V   W   X   Y   Z
```

FIGURE 14–2 Help Screen

MENUS

Whether they are "pull-down," "pop-up," or menu "bars" spread across the top of the screen, computer menus list the choices available to the user. Usually, each screen has a set of menu options displayed in a "bar" that remains constantly visible, even as the rest of the screen's content changes (see Figure 14-3). To use this bar, users simply position the cursor

FIGURE 14-3 Menu Bar

Menu Bar
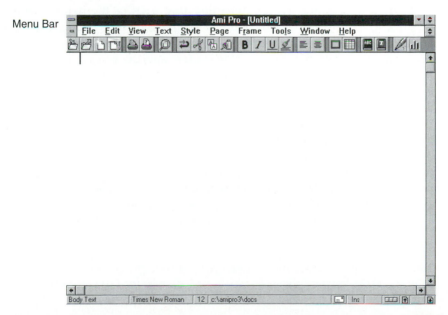

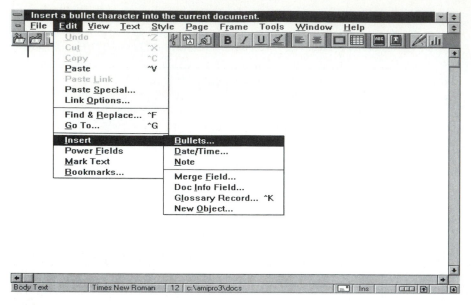

FIGURE 14-4 Layered Pull-Down Menu

over the menu item they want and either press "Enter" or click the mouse button, depending on the interface they are using. For example, suppose you are working with a graphics software package and want to know the options for designing text. Using a mouse, you can move the cursor to the top of the screen where the menu options are displayed, move the cursor over the word "text," click, and a list of the various available options appears in a drop-down ("pull-down") box. In some applications, these pull-down menus are layered, meaning that each of the items in the first menu also has a sub-menu of options. To make choices from these various menu levels, you move the cursor again to click on the option you want, the program automatically registers this choice, and the box disappears. If you decide not to make a choice at that time, you should be able to escape the menu easily by clicking on an escape key or a specially designated icon, and the box disappears. Figure 14-4 shows a layered pull-down menu.

Designing effective menus requires creativity and a careful attention to organizational detail. The foremost design rule is that the menus should be hierarchical in structure. In other words, the menus ought to have a distinct order in which each menu category begins with the broader topics and proceeds to narrower and narrower subcategories of information. For example, you may click on the beginning of a menu labeled "Edit" and be shown a pull-down menu that includes the term "Text." Clicking on the sub-menu "Text," you are given the list: Fonts, Sizes, Styles, Justification, Text Properties. If you click on "Sizes," you see another subcategory: Plain Text, Bold, Italic, Underline, Superscript, Subscript. The hierarchy continues to move to narrower choices as you move deeper into the menu structure.

A second important rule about menu design is that the terms used in the menus should be quickly understandable, consistent throughout the document, and concise. Terms such as "Text mode only" may have no meaning for the users, especially if the same concept is labelled as "Write mode" in another location in the document. Pay attention to the users perceptions and try to design menus that are clearly logical and use terms the users will understand.

DIALOGUE BOXES

A dialogue box is a message that appears on the screen and asks the user questions. Users may answer yes/no or be given several options from which to choose. If the box gives options, it is functioning in much the same way a pull-down menu does. Some software documentation refers to these lists of choices as menus and some refer to them as dialogue boxes—the distinction is blurred. For instance, a dialogue box may ask a straightforward question: "Do you want to save the text?" to which users can answer "yes" or "no." But if the box presents a list of possible formats, users respond to it as they would to a menu: by either typing their response or clicking on the appropriate choice. Figure 14-5 illustrates various kinds of dialogue boxes.

FIGURE 14-5 Dialogue Boxes

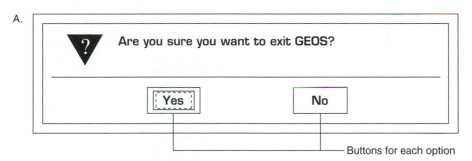

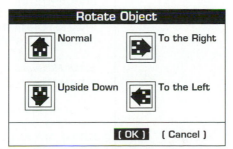

ERROR MESSAGES AND STATUS MESSAGES

In every software program, users will eventually see error messages. Often accompanied by a "beep" or some other audio warning, the messages let users know that they have made a mistake and must try again. These messages may appear on the command line at the top of the screen or wherever they are most likely to get attention. For example, if you attempt to create a file with the same name as one that already is present in the directory, you may get the message: "File already exists," and you know you must find a different name for the current file—or perhaps you are repeating work you have already done.

When writing error messages, pay particular attention to tone. Too often, these messages create negative impressions for users by emphasizing what is incorrect rather than ways to fix the problem. For example, a message such as "Bad string," unnecessarily makes a value judgment on the user's actions. Rewriting the message to say "Check string" alerts users to the problem and avoids the negative connotations. It is also a good idea to provide a clue for the users about where in the documentation they can go for help in correcting the difficulty. The message, "Check string. See C-2," not only has a more positive tone, it also points users to a specific explanation of the problem.

Unlike error messages, status messages do not warn users about their mistakes; instead, they keep them informed about what the software is doing. A status message may say: "Spell checker working" or "Formatting disk in drive A." These mini-reports tell you that the software is doing the job you have asked it to do. See Figure 14-6 for examples of both status and error messages.

▬▬ **FIGURE 14-6** Status and Error
    Messages

Error Messages:
NONSYSTEM DISK OR DISK ERROR
PRINTER ERROR
ERROR WRITING
CAN'T CREATE TEMP FILE
INVALID FORMATTING COMMAND
FILE OPEN QUIT ANYWAY?

Status Messages:
FORMATTING....
DONE
RESTORE DEFINE
WORKING....
STRIKE ANY KEY WHEN READY

ONLINE TUTORIALS

An online tutorial is software that functions as a teacher. For example, when users buy some new software or hardware, they may also receive a tutorial disk whose purpose is to teach them how to use the product. As in hardcopy tutorials, these online versions should provide short, step-by-step lessons that offer sufficient feedback so that users can gain confidence in their learning progress as they go along. Often these tutorials contain audio components and quite a few graphic elements to make the learning process interesting and memorable. If the disk is a sales demo, the "tutorial" will have many more of these marketing features than it would have if it were primarily a teaching tool.

It is worth noting here that while planning these tutorials, technical writers can have a major impact on the software design. As you are preparing to write the online material, you have the opportunity to test the product's user interface. Does it work well? Will users find the product easy to use? In writing the tutorial, you are looking at the product from the users' perspective rather than from that of the engineers. By commenting on the design of the software at this stage and working closely with the development engineers, you can help them produce a better product.

HYPERTEXT

Some technical communicators believe that hypertext is *the* answer to the problem of managing knowledge. While that view may be extreme, most people agree that hypertext is an important step in the advancement of communication. But what is it, exactly? According to an article by Christine Borgman and Bruce Henstell:

> Hypertext is an application, or a technique, wherein units of textual data can be more or less effortlessly linked. Linked, they form clusters or nodes, perhaps stacks. Both nodes and links can be given attributes. When graphics, video, or digitized sounds are introduced, hypertext becomes "hypermedia." The hypertext database is a "hyperdocument." The links and nodes in a hyperdocument may be predefined (as in an information retrieval application), or they may be formed by the user (as in a writing application).

> "Hypertext: What's in a Name?" *Bulletin of ASIS,* 1989.

What this technique means for technical communicators is that computers can now make it incredibly easy for users to manipulate information in ways that suit individual needs. A hypertext program produces stacks of "cards" (screens), each containing information and a "button" (a link to other information). The information on the cards can be text, graphics, or both. The buttons can point to cards, stacks, or other sources. These sources can be graphics, voice, music, animation, or video. Hypertext programmers can create stacks that link all sorts of information and that allow users to move from card to card in the order that works best for them. There is no one set path through the hyperdocument, though the number of paths may be finite depending on the document's design.

For example, using hypertext, teachers can create stacks on particular topics and allow students to explore these subjects on their own and at their own pace. Customers at a suburban shopping mall can use a hypertext stack rather than the mall directory to find just the right merchandise and prices to fit their needs. Filmmakers can link videotaped sequences and play them back in any order without using splicing equipment. The possibilities for applications are seemingly endless.

Technical writers are currently developing hypertext stacks to explain products or entire technologies for both corporate clients and company employees. New hires in any department find hypertext extremely useful in learning the ropes during the first few weeks, and seasoned employees such as maintenance engineers, field personnel—even programmers—can use hypertext applications to access information on a variety of specialized topics. Figure 14-7 gives a hardcopy sample of a few hypertext cards.

=== EXERCISE ===

List all the online documentation types you can find in one software package. Do not limit yourself to the types listed in this section—add more if you find them. After you have generated the list, go back and look at each type to determine if it is easy to use and if it is really useful. Explain your decisions. Would you make any changes in the way the online material appears? Why or why not?

THE BASICS OF SCREEN DESIGN

Designing screens is a relatively simple process if you follow a few standard rules:

1. Keep the text concise.
2. Avoid multiple screens for one piece of information.
3. Avoid dense blocks of text.
4. Use consistent placement.
5. Always indicate escape and navigational methods.
6. Avoid the "circus effect."

Keeping the text concise means that you should use lists instead of prosy paragraphs if possible, and even when prose is necessary keep it to a minimum. Using graphics and icons effectively can help eliminate excess words.

To avoid multiple screens you should think of the information you must provide as "chunked" into one-screen modules. Much of the guidelines given in Chapter 6 for modular documentation apply to screen design as well. If the user must scroll beyond one screen to complete the information, your design is not as efficient as it could be.

Avoid dense blocks of text for several reasons. First, such text is formidable to read in any situation; and second, prose is much harder to read on screen than on paper, and dense prose makes the potential eye strain for users even worse.

Consistently placing information elements in the same place on every screen creates user expectations that make your documentation more readable. The users quickly come to expect the menus to be in one spot, the command line in another, and the escape information in yet another. It may help for you to design a grid template that you use for every screen so that the information is located *exactly* in the same place each time. Familiarity with screen terrain gives users a greater sense of confidence and increases their ability to use your documentation productively.

Always indicate escape and navigational methods on each screen. Users may not read or remember the preliminary instructions given for moving around in the documentation. If every screen has a clear signal as to how to continue in the online information and how to exit the material, users will never feel lost or abandoned in a morass of on-screen text. They will always feel in control.

Avoid the "circus effect" created by using too many colors, sounds, or other attention-getting devices. When too many of these devices compete for users' attention, the result is similar to a three-ring circus where no one can possibly focus clearly on the main event. Use visual and audio aids sparingly and make sure that they are enhancing the point you are making, not competing with it for attention.

15/29/909:19 AMDevice Configuration Guide

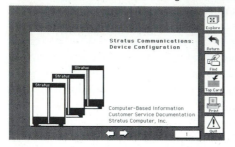

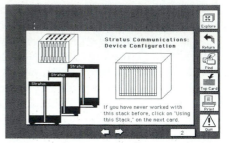

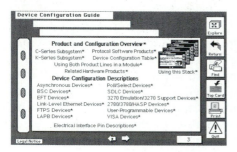

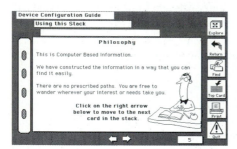

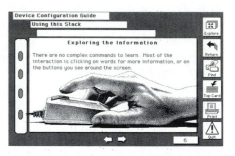

FIGURE 14-7   Hypercard Stack (Used by permission of David Balzoti, Stratus Computer, Inc.)

25/29/909:19 AMDevice Configuration Guide

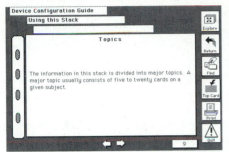

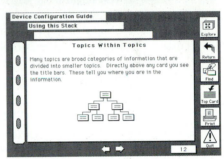

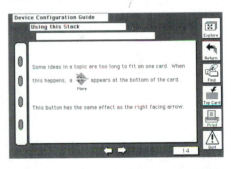

1. Design a series of help screens for a software program. Carefully draw on paper
    the design for at least five related screens (one page for each screen), remember-
    ing to follow the rules given in this section.

2. Analyze the online documentation for a software package. Does it contain all
    the online material it should? Is the material user-friendly? Is it designed well?
    Present your findings in written form.

3. If you have access to authoring software, design a short online tutorial for a few
    features of a software product. After you have finished the tutorial, make a list of
    suggestions you might make to the developers to help them improve the soft-
    ware's interface design.

4. If you have access to a hypertext authoring system (Apple's Hypercard, for
    example), create a hypertext stack that gives the public information on a particu-
    lar topic. Some examples are listed below:

    • Restaurant options in your area.

    • Merchandise available in a local department store.

    • Style guide for basic punctuation.

    • Guide to a local organization (including officers, committees, and so on).

    • Information on a new product or service.

DOCUMENTATION EVALUATION GUIDE

Adapted from the "Judging Form," Boston Chapter, Society
for Technical Communication, Technical Publications Competitions.

JUDGING FORM
STC PUBLICATIONS COMPETITIONS

Summary of Evaluation
(Please Complete This Section Last)

Major Strengths of Entry:

_____

_____

_____

_____

Major Weaknesses of Entry:

_____

_____

_____

_____

	Poor			Excellent	
Summary Rating:	1	2	3	4	5

		Poor			Excellent	
1. AUDIENCE AND PURPOSE	Rating:	1	2	3	4	5

   a. Is the purpose clearly stated?
   b. Does the document fulfill the purpose?
   c. Is the audience clearly defined?
   d. Does the document meet the audience's needs?

		Poor			Excellent	
2. ORGANIZATION	Rating:	1	2	3	4	5

   a. Is the organization appropriate for the purpose?
   b. Is the organization logical at all levels?
      (Overall? Subsequent? Paragraphs?)
   c. Can users find what they need?
   d. Is the Table of Contents complete and useful?
   e. Are the headings helpful?
   f. Is the index comprehensive, easy to use, and cross-referenced?
   g. If headers and footers are used, are they helpful?
   h. Are references and footnotes relevant and clear?

		Poor			Excellent	

3. **CONTENT** Rating:

Poor   1   2   3   Excellent   4   5

  a. Is the content relevant to the purpose?
  b. Is the content geared to the stated audience?
  c. Are the main points properly stressed?
  d. Are there sufficient helpful examples?
  e. Are there unnecessary digressions?

4. **WRITING AND EDITING** Rating:

Poor   1   2   3   Excellent   4   5

  a. Is the reading level appropriate to the audience?
  b. Are the tone and style appropriate for the purpose and audience?
  c. Is terminology consistent?
  d. Are grammar, syntax, spelling and punctuation correct?
  e. Is the passive voice used only when necessary?

5. **ILLUSTRATIONS** Rating:

Poor   1   2   3   Excellent   4   5

  a. Do the illustrations contribute to the usefulness of the document?
  b. Are the illustrations effectively integrated into the text? Are they placed appropriately?
  c. Are the illustrations clearly labeled?
  d. Are the illustrations presented neatly and consistently?

6. **LAYOUT AND DESIGN** Rating:

Poor   1   2   3   Excellent   4   5

  a. Is the layout effective for the audience and purpose?
  b. Is the overall design consistent and coherent?
  c. Do typefaces and point sizes promote readability?
  d. Does the use of color add to the effectiveness?
  e. Is white space used effectively?
  f. Are there enough headings with the text to aid in information retrieval?

PLEASE COMPLETE THE SUMMARY OF EVALUATION ON PAGE 1

INDEX